PENGUIN BOOKS

HE KNEW HE WAS RIGHT

'There is no doubting [Lovelock's] simple goodness and honesty, nor is there any question about his natural scientific genius. These qualities shine through his authorized biography by science writers, John and Mary Gribbin. The impression they give is of a charming, humorous, modest fellow with whom you could happily discuss any topic under the sun, whether you agreed with each other or not' John Michell, *Spectator*

'Wonderfully lucid' Jonathan Bate, *Sunday Telegraph*

'Gives a good sense of Lovelock's inspirational, independent spirit' Roger Highfield, *Daily Telegraph*

'An absorbing new biography' Michael McCarthy, *Independent*

'Demonstrates well how Gaia has overcome its main critics to become part of a distinguished historical tradition of serious if controversial science' *New Scientist*

'James Lovelock is one of the great thinkers of our time. His ideas and inventions have opened up new insights into our planet and the way it works, and the story behind them will appeal to a very wide audience. I am pleased to recommend this book' Chris Rapley, director of the Science Museum, London

ABOUT THE AUTHORS

John and Mary Gribbin are two of today's greatest writers of popular science. Together they have collaborated on many books, including *Ice Age*, *Richard Feynman: A Life in Science* and *Stardust*. John is also the author of bestselling titles including *In Search of Schrödinger's Cat*, *Science: A History* and *Deep Simplicity*.

JOHN GRIBBIN AND
MARY GRIBBIN

He Knew He Was Right

The Irrepressible Life of James Lovelock

PENGUIN BOOKS

PENGUIN BOOKS

Published by the Penguin Group
Penguin Books Ltd, 80 Strand, London WC2R ORL, England
Penguin Group (USA), Inc., 375 Hudson Street, New York, New York 10014, USA
Penguin Group (Canada), 90 Eglinton Avenue East, Suite 700, Toronto, Ontario, Canada M4P 2Y3
(a division of Pearson Penguin Canada Inc.)
Penguin Ireland, 25 St Stephen's Green, Dublin 2, Ireland
(a division of Penguin Books Ltd)
Penguin Group (Australia), 250 Camberwell Road, Camberwell, Victoria 3124, Australia
(a division of Pearson Australia Group Pty Ltd)
Penguin Books India Pvt Ltd, 11 Community Centre, Panchsheel Park, New Delhi – 110 017, India
Penguin Group (NZ), 67 Apollo Drive, Rosedale, North Shore 0632, New Zealand
(a division of Pearson New Zealand Ltd)
Penguin Books (South Africa) (Pty) Ltd, 24 Sturdee Avenue, Rosebank, Johannesburg 2196, South Africa

Penguin Books Ltd, Registered Offices: 80 Strand, London WC2R ORL, England

www.penguin.com

First published by Allen Lane 2009
Published in Penguin Books 2009
1

Printed in Great Britain by Clays Ltd, St Ives plc

A CIP catalogue record for this book is available from the British Library

978-0-141-03161-3

www.greenpenguin.co.uk

Contents

List of Illustrations

Citation for the Wollaston Medal, awarded to James Lovelock by the Geological Society in 2006

Even in the illustrious history of the Society's senior medal, first awarded to William Smith in 1831, it is rare to be able to say that the recipient has opened up a whole new field of Earth science study. But that is the case with this year's winner, James Lovelock.

Lovelock does not lack for honours after his long and distinguished career in science. As well as more lately being created Companion of Honour and Commander of the Order of the British Empire, he became a Fellow of the Royal Society in 1974, and garnered many awards for his pioneering work in chromatography. Lovelock invented the electron capture detector for gas chromatography – an instrument whose exquisite sensitivity has subsequently been central to several important environmental breakthroughs. For example, during the 1960s it enabled the documentation of widespread dissemination of harmful and persistent pesticides like DDT, and later on the technique was extended to the polychlorinated biphenyls (PCBs). Lovelock himself famously used the technique to chart the ubiquitous presence of chlorofluorocarbons – CFCs – in the atmosphere, triggering the discoveries (by Rowland and Molina) of the harmful influence of CFCs on atmospheric ozone – work for which they received the Nobel Prize for Chemistry in 1997. He has also developed instruments for exploring other planets than our own, including those aboard the two Viking craft that went to Mars in 1975, about which I know he will tell us in a moment.

But Lovelock really came to high public prominence for the scientific concept that has captured the imaginations of Earth scientists, biologists and public alike – the concept for which we as geologists chiefly honour him today – the Gaia Hypothesis and Theory. This view of the planet and the life that lives on it as a single complex system, in some ways analogous to a homeostatically self-regulating organism, is what has given rise to the field we now know as 'Earth System Science', also the most recently formed of this Society's Specialist Groups.

It is hard to overemphasize the unifying nature of this holistic worldview, which has broken down artificial disciplinary barriers that have existed since the late eighteenth and early nineteenth century when Societies such as this were first formed, and the wonderful richness of insight that has flowed from the multidisciplinarity that has followed. This is especially so in the understanding of feedback loops between life and the environment, especially the dimethyl sulphide – cloud albedo – surface temperature (CLAW) hypothesis, and the whole idea that life coupled with its material environment regulates planetary temperature and chemical composition over long timescales by influencing rates of silicate weathering.

James Lovelock, it gives me enormous pleasure to reward this towering career with the highest honour the Society can bestow.

Preface

The Edge of Chaos

In Greek mythology, Chaos was the first thing to appear, a formless void out of which Gaia, the Mother of the Earth, emerges. In modern science, chaos theory describes the behaviour of systems which are so sensitive to conditions that a very small change in one property can lead to a large and unpredictable change in the whole system. As we described in our book *Deep Simplicity*, stable systems which are insensitive to their conditions are dull and uninteresting places where nothing changes. At the other extreme, completely chaotic systems are so wild and unpredictable that nothing interesting can happen there, either. But right on the edge of chaos, where stability begins to break down, interesting things happen and complex structures emerge.

To highlight how this happens, we can imagine a smoothly flowing river in which there is one large rock sticking out above the surface. The flowing water of the river divides around the rock, and joins up again seamlessly on the other side, so that small chips of wood floating on the water follow these 'streamlines'. If there is rain upstream, the flow of the river will increase, and as it does so it goes through at least three distinct changes.

First, as the flow builds up, little whirlpools appear behind the rock. These vortices stay in the same place, and a chip of wood floating downstream may get trapped in one of them and go round and round for a long time.

At the next stage, as the speed of the water flowing downstream increases, vortices form behind the rock, but don't stay there. They detach themselves, and move away downstream, retaining their separate existence for a while before dissolving into the flow. As they do so, new vortices form behind the rock and detach in their turn. Now, a chip of wood might get caught in one of these eddies and be carried downstream still circling around in the vortex as long as the vortex lasts.

As the flow of water increases still further, the region behind the rock where the vortices survive gets smaller and smaller, with vortices forming and breaking up almost immediately to produce a choppy surface in which there are seemingly only irregular fluctuations – turbulence. Eventually, when the flow is fast enough all trace of order in the region behind the rock disappears. No vortices form and the entire surface of the water breaks up behind the rock into unpredictable chaotic motion.

Interesting things happen on the edge of chaos, and one of those interesting things is life – in particular, life on Earth. But the edge of chaos is a dangerous place to live, because small changes in the environment can, in certain circumstances, lead to large and not always predictable consequences. Humankind is now making just such a change to the environment of the Earth, causing a rapid (by geological standards) rise in temperature which, it is claimed, may flip the climate system into a new state, like a vortex breaking off and floating away downstream. This emergence of chaos in the scientific sense could produce chaos in the everyday sense of the word, as society collapses under the pressure.

Gaia is also the name given to a theory which describes how the different components of the Earth System, living and non-living, have worked together for eons to maintain conditions suitable for life. In a reversal of the story of the Greek myth, could chaos be about to bring about the death of Gaia? Are we also living on the edge of chaos in the everyday sense of the term? To answer that question we will need to investigate the relationship between global warming and Gaia, encapsulated in the work over the past forty years of James Lovelock, the father of Gaia theory.

Introduction

'I don't understand what's going on. I'm delighted, but I'm baffled. Absolutely baffled. It's not the sort of thing you expect at the age of 87.' Jim Lovelock had just returned from a lecture tour of the United States, where he had been promoting his book *Revenge of Gaia*. For half his life he had been promoting the idea that the living and non-living components of Planet Earth act as components of a single system, dubbed 'Gaia' by the novelist William Golding, in which biological processes play a major part in controlling the physical environment. For almost all that time, the idea – which crucially sees the Earth System involving both living and non-living components in feedbacks which have, until now, maintained conditions suitable for life – had been ridiculed by the scientific establishment. Even when some of Lovelock's ideas began to become part of the mainstream, most scientists couldn't bring themselves to use the term 'Gaia', preferring to refer to Earth System Science. But in 2006, everything changed.

On his tour, Lovelock spoke at universities across the USA. 'Everywhere I went,' he told us in amazement, 'I was introduced as "the originator of Gaia theory", which they described as "the basis of Earth System Science".' Still slightly jet-lagged from the trip, he shook his head in wonder. 'They didn't even call it a hypothesis. They called it "Gaia", and "theory". Something's happened this year to cause a sudden swing from "this is dubious, we're not sure about it," to "this is solid science."' His American wife, Sandy, proudly took up the tale: 'And what Jim is too modest to tell you is that everywhere we went he received standing ovations. Sometimes, even before he spoke.'

We were visiting Jim and Sandy at their remote home on the Devon/Cornwall borders, to lay the groundwork for this book; not just a biography of Lovelock himself, but also an account of the intertwined stories of Gaia theory and global warming – for, of course, it was the sudden appreciation that the world really is warming at an unprecedented rate, and that this is due to the activities of humankind, that had equally suddenly made Gaia theory respectable. It is now widely recognized that the most critical global problem facing us today is environmental. Whether or not climate change itself has reached a 'tipping point', after which drastic change is inevitable, the middle of the first decade of the twenty-first century seems to have marked a tipping point in the public perception of climate change, which is now seen as a real and dramatic threat. Lovelock has been warning about the danger of global warming for decades, but had generally been regarded as a somewhat hysterical prophet of doom. In 2006, it became clear that he had been right all along.

After coffee, we all went for a walk around the thirty-five acres of land that the Lovelocks are 'giving back to Gaia'. Although it was already the last week in September, the temperature was in the low twenties; we came to a wide river bed, where a trickle of water flowed across the stones. 'That's the lowest I've seen it since I moved here thirty years ago,' said Jim. 'Hard to believe there used to be salmon in the river here.' But on a brighter note, he told us that in the unmanaged woodlands wild orchids have returned, their seeds carried on the wind; and we saw for ourselves several badger setts. In the meadows, wild flowers proliferate in spring.

After our walk, we got down to work, pulling out the material we needed from the Lovelocks' archive, pausing now and again to linger a little over some key document, and to note how attitudes had changed. Where a generation ago Gaia was regarded as at best cranky and at worst alarmist, it is now more and more accepted as a warning truth, an idea whose time has come and an idea we must embrace to save the world.

Even a quick glance at the material confirmed our view that the time is therefore ripe for a book which sets Lovelock's life and work in the context of what is happening to our planet now, and which spells out for those who still think of Gaia as some kind of crazy left-field idea that the oneness and mutual interdependence of both life and non-life is the key to understanding both the past and the future of the Earth. We have known Lovelock for more than thirty years, but the specific trigger for this book, and the reason for our visit to him in September 2006, came at an event earlier that year in Brighton, where we were struck by the response of young people to him and his ideas. Few of those people who now see Lovelock as an iconic figure know about his life and his earlier work as a chemist and inventor – work which included twenty years of medical research (among other things, discovering just how the common cold is spread by coughs and sneezes), inventing the detectors used to search for life on Mars, providing scientific backup for the emotional message of Rachel Carson's book *Silent Spring*, and the observations which led to the discovery of the depletion of ozone in the now-notorious 'hole' over Antarctica.

In his personal life, he was a Quaker and conscientious objector in the Second World War (later changing his mind in view of the evils of Nazism, but continuing to see the policy of mass bombing of civilian targets as a crime), supported his family for a time on a visit to the United States by selling his blood, and gave up a salary of $40,000 a year (a small fortune in 1963) to become an independent scientist based in an English village – a base from which all his best-known work emerged.

After such a long and exciting life, as he entered his ninth decade at the beginning of the twenty-first century, Lovelock was contemplating a quiet retirement and wondering what the next decade would hold – he did not anticipate that just over halfway through that decade his profile would be higher than ever, he would be a bestselling author, and he would be embroiled more deeply than ever in the fight to get politicians to see sense about global warming. But Lovelock is no ordinary octogenarian. Two days earlier, he had heard that his contract as a consultant to the Ministry of Defence had just been renewed, giving him, as he gleefully pointed out, the prospect of maintaining the relationship up to the age of 90. Shortly after our meeting that September day in 2006 he was off again, this time to Japan (via Cardiff) to spread the word about Gaia. We returned to Sussex with a car full of Lovelock's papers and several hours of recorded reminiscences, to begin to try to do justice to one of the most important scientists of the past hundred years.

This book is written not in the expectation of making any converts to the acceptance of the reality of global warming and Gaia theory, both of which are now well established and need no proselytizing; we do hope to set in its historical context the development of one of the crucial ideas of the twentieth century, and the way it was developed by a most unusual scientist, one whose entire life can be seen, with hindsight, as a unique preparation for his revolutionary insight. But the story begins more than a hundred years before he was born.

John Gribbin
Mary Gribbin
March 2008

Prologue

A Beginner's Guide to Gaia

The Gaia concept was first formulated as a scientific hypothesis by James Lovelock, in the 1960s. It grew out of his work for NASA on experiments designed to detect life on Mars. The idea was presented in scientific papers in the 1970s and in a book, *Gaia: A New Look at Life on Earth*, in 1979. The essence of Lovelock's argument is that the Earth can be regarded as a single living system, in which components traditionally regarded as 'non-living' (such as the cycle of weathering of rocks) are important to components traditionally regarded as alive, while components traditionally regarded as alive are important to 'non-living' systems. Feedbacks involving both kinds of process have, according to the Gaia hypothesis, maintained conditions suitable for life on our planet over billions of years, in spite of external threats such as the steady warming of the Sun which would otherwise have made the Earth uninhabitable. A good analogy, made by the American Lewis Thomas, is that the Earth is like a single living cell; or, as a student of the biologist Lynn Margulis put it, 'symbiosis seen from space'.

A *hypothesis* is a scientific idea put forward to explain known facts but as yet untested by experiment. A *theory* is a scientific idea that has made successful predictions about the outcome of experiments, and is therefore much stronger than a hypothesis. The Gaia concept is now widely regarded as a fully fledged theory, because Lovelock and others have used it to make predictions which have been borne out by experiment and observation – in particular, the idea of Daisyworld, which we discuss in Chapter 8. The original hypothesis was proposed to explain why active chemical substances such as oxygen and methane persist in stable concentrations in the Earth's atmosphere, making Earth the Goldilocks Planet – 'just right' for life. Lovelock suggested that the long-lasting presence of these gases is a signature of life, and that detecting such substances in the atmospheres of other planets would be a reliable way to detect life. The corollary was that since the atmospheres of Mars and Venus contain scarcely anything except the stable, unreactive gas carbon dioxide, there can be no life on either of them. This prediction has (so far) been borne out by observations made by spaceprobes, and is another successful prediction of Gaia theory.

Lovelock's idea received little attention from the scientific community until he wrote an article about it in the magazine *New Scientist* in 1975. This led in 1979 to his first book on the subject; it provoked a hostile response from many evolutionary biologists at the same time that it was taken up and championed with almost religious fervour by some environmentalists and members of the Green movement. In both cases, the choice of the name Gaia (suggested to Lovelock by his friend and neighbour the novelist William Golding from that of the Greek goddess of the Earth) played a big part in determining the strength of the reaction.

The chief criticism of Lovelock's idea was that as the Earth is a single system, there is no way for competition between different planets to produce the kind of world we live in by natural selection. How could different individuals in living populations, let alone the non-living components of the system, work together for the good of all when each individual has evolved through selection and 'survival of the fittest'? Richard Dawkins, best known as the proponent of the idea of the selfish gene, argued in his book *The Extended Phenotype* (1982) that this would require foresight and planning. He dismissed the idea that feedback could stabilize the Earth System, and said 'there is no way for evolution by natural selection to lead to altruism on a global scale'. Lovelock addressed these criticisms in his later work, in particular his book *The Ages of Gaia* (1988). Apart from specific rebuttal of some of the criticism, he argues that critics such as Dawkins take too narrow a view, while he sees things from the perspective of a general practioner rather than a specialist – the medical analogy is apt, because Lovelock cut his scientific teeth in medical research.

A related criticism is that the planet as a whole cannot be alive, because it cannot reproduce. But Lovelock points out that this rests upon a very limited definition of life centred on the ability to replicate and pass on genetic information to succeeding generations. It would exclude as 'non-living' such obviously living things as a mule, or a post-menopausal woman. A better definition of life involves self-sustaining systems occurring in feedback loops, feeding off an external flow of energy, as we discuss in Chapter 3.

Lynn Margulis sees things slightly differently from Lovelock, although she is an enthusiastic supporter of the idea of Gaia. She says that Gaia is 'not an organism', but 'an emergent property of interaction among organisms'. Gaia is 'the series of interacting ecosystems that compose a single huge ecosystem at the Earth's surface.' But she agrees with Lovelock that, 'the surface of the planet behaves as a physiological system in certain limited ways.'

The debate about Gaia has intensified in the twenty-first century because of its relevance to the problem of global warming caused by a buildup of greenhouse gases in the atmosphere as a result of human activities. Lovelock believes that unless drastic action is taken to alleviate the situation, the increase in temperatures will soon reach a tipping point that takes the Earth System beyond the limits where the present set of natural feedbacks can operate, and that before the end of the present century the planet will 'flip' into a much hotter state maintained by a different set of feedbacks. This conflicts with the conventional wisdom, which assumes that, although the consequences of the warming may be unpleasant, it will proceed in a more or less steady fashion for decades, with no sudden jumps.

The increasing pace of the observed climate change has stimulated research in all areas of Earth System Science, and the shift of opinion in favour of Gaia theory is highlighted by the 'Amsterdam Declaration' made jointly in 2001 by four international global change research programmes – the International Geosphere-Biosphere Programme (IGBP), the International Human Dimensions Programme on Global Environmental Change (IHDP), the World Climate Research Programme (WCRP) and the international biodiversity programme DIVERSITAS. They 'recognize that, in addition to the threat of significant climate change, there is growing concern over the ever-increasing human modification of other aspects of the global environment and the consequent implications for human well-being', and go on to say:

Research carried out over the past decade under the auspices of the four programmes to address these concerns has shown that:

- The Earth System behaves as a single, self-regulating system comprised of physical, chemical, biological and human components. The interactions and feedbacks between the component parts are complex and exhibit multi-scale temporal and spatial variability. The understanding of the natural dynamics of the Earth System has advanced greatly in recent years and provides a sound basis for evaluating the effects and consequences of human-driven change.
- Human activities are significantly influencing Earth's environment in many ways in addition to greenhouse gas emissions and climate change. Anthropogenic changes to Earth's land surface, oceans, coasts and atmosphere and to biological diversity, the water cycle and biogeochemical cycles are clearly identifiable beyond natural variability. They are equal to some of the great forces of nature in their extent and impact. Many are accelerating. Global change is real and is happening now.
- Global change cannot be understood in terms of a simple cause–effect paradigm. Human-driven changes cause multiple effects that cascade through the Earth System in complex ways. These effects interact with each other and with local- and regional-scale changes in multidimensional patterns that are difficult to understand and even more difficult to predict. Surprises abound.
- Earth System dynamics are characterised by critical thresholds and abrupt changes. Human activities could inadvertently trigger such changes with severe consequences for Earth's environment and inhabitants. The Earth System has operated in different states over the last half million years, with abrupt transitions (a decade or less) sometimes occurring between them. Human activities have the potential to switch the Earth System to alternative modes of operation that may prove irreversible and less hospitable to humans and other life. The probability of a human-driven abrupt change in Earth's environment has yet to be quantified but is not negligible.
- In terms of some key environmental parameters, the Earth System has moved well outside the range of the natural variability exhibited over the last half million years at least. The nature of changes now occurring simultaneously in the Earth System, their magnitudes and rates of change are unprecedented. The Earth is currently operating in a no-analogue state.

This is as good a summary as any of why Gaia theory is so important today.

I

The Greenhouse before Gaia

Scientific understanding of human-induced global warming is older than you might think. The idea that carbon dioxide released into the atmosphere by burning fossil fuel would warm the planet was clear to at least a few scientists more than a hundred years ago; and that was barely two hundred years after the scientific revolution which, among other things, led to an understanding of atmospheric chemistry. The importance of what is now called the anthropogenic greenhouse effect began to emerge not in the late twentieth century through the work of people like James Lovelock, but in the early nineteenth century through the work of people like Jean-Baptiste Joseph Fourier (usually known as Joseph Fourier).

Of course, nobody could begin to understand the role of carbon dioxide in keeping our planet warm until it was known what carbon dioxide was, and that it was present in the atmosphere. In the seventeenth century, Robert Boyle had begun to appreciate the nature of the atmosphere when he described it as the product of the 'exhalations of the terraqueous globe', a rather Gaian description by which he meant the products of volcanic activity, decaying vegetation and animal life. Although this seems obvious today, it was a profound step forward from the old idea that the atmosphere was made up of some mystical substance known as the ether. It was only in the 1750s that Joseph Black showed that air is a mixture of gases, not a single substance, and isolated one of those gases, then known as 'fixed air' but now called carbon dioxide – the first component of the atmosphere to be identified. Two decades later, Daniel Rutherford isolated nitrogen from air, and oxygen was identified by Joseph Priestley and independently by Carl Scheele. In the early 1780s, Henry Cavendish determined the composition of air to be almost exactly 79 per cent nitrogen and 21 per cent oxygen, with just traces of other gases, including carbon dioxide; the scene was set for nineteenth-century scientists to begin to understand how this blanket of air keeps the Earth warm.

Although nobody at the time had any inkling of the role that his discovery would play in the story of global warming, with hindsight that story can be seen to begin in 1800, when the astronomer William Herschel was studying the warming effect of light from the Sun passing through different prisms and coloured filters. To his surprise, he found that when light from the Sun was split up into a rainbow pattern by a prism, a thermometer placed beyond the red end of the spectrum warmed up, even though it was receiving no visible light from the Sun. He had discovered what later became known as 'infrared' radiation – radiation like light but with wavelengths longer than red light, invisible to our eyes. But it was a quarter of a century before this invisible radiation was first linked with global warming.

Fourier, who was born in 1768 and died in 1830, came to the study of global warming late in his life, but he is the first person known to have appreciated that the atmosphere keeps the Earth warmer than it would be if it were a bare ball of rock orbiting at the same distance from the Sun. Fourier was very interested in the way heat is transmitted, and among other things he calculated an estimate for the age of the Earth based on how long it would have taken for a ball of molten rock to cool to the Earth's present state. His estimate was a hundred million years, a number so staggeringly large that he didn't dare publish it – many people in his day still believed in the age of the Earth based on a literal interpretation of the chronology in the Bible, which comes out at about six thousand years. But Fourier's estimate is only 2 per cent of the best modern estimates for the age of the Earth.

The calculation of how hot (or rather, how cold) a bare ball of rock orbiting the Sun at the same distance as the Earth would be is relatively straightforward, and Fourier and his contemporaries got it more or less right. But we don't have to worry too much about the calculation because there is indeed a bare ball of rock orbiting the Sun at the same distance as the Earth – our Moon – and scientists have measured its average temperature. Like the Earth, the surface of the Moon gets colder at night and warmer by day, but averaging over the whole ball of rock it is a chilly −18°C. Averaging over the entire Earth in the same way, the surface temperature is 15°C. Something keeps the surface of the Earth about 33°C warmer than it would otherwise be – and that something, as Fourier realized, is the atmosphere. He carried out his studies of global warming in the 1820s, and in 1824, summarizing work that he had previously reported in various places, he wrote that 'the temperature [of the Earth] can be augmented by the interposition of the atmosphere, because heat in the state of light finds less resistance in penetrating the air, than in repassing into the air when converted into non-luminous heat'.[1] In other papers published in that decade he made the analogy that heat is trapped near the surface of the Earth by the atmosphere in the way that heat is trapped inside a hothouse. Specifically, he referred to the warming inside a box with a glass cover, exposed to the Sun, and suggested that the glass lid retained the 'obscure radiation' (now known as infrared) inside the box. His analogy was wrong, but much later the term 'greenhouse effect' came to be almost inextricably associated with global warming – so much so that we shall use it in this way ourselves.

Why was Fourier wrong? The air in a greenhouse gets hot because the rays from the Sun passing through the panes of glass in the greenhouse heat the ground inside the greenhouse, which gives up warmth to the air above it. Hot air rises, and outside a greenhouse air warmed in this way rises and carries the heat with it, eventually radiating it away into space. But inside the greenhouse the warm air cannot escape and the air gets hotter and hotter. Greenhouses get hot because their roofs suppress convection, which is why gardeners adjust the temperature inside by opening or closing vents in the roof.

The first scientist to appreciate the real role of the atmosphere in warming the globe was John Tyndall (1820–1893), an Irish polymath who was also one of the first popularizers of science and whose lectures in the United States were almost as celebrated as those of his contemporary Charles Dickens.

1. An English translation of the 1824 paper was published in the *American Journal of Science* in 1837, so the work was widely known.

Tyndall, the son of a local policeman, was born in the village of Leighlin Bridge, in Carlow. He received only a basic formal education, but in 1839 he got a job with the Irish Ordnance Survey, and in 1844 he became a railway engineer with a company based in Manchester. All the while, he had been studying and attending lectures in his spare time, and in 1847 he was appointed as a teacher of mathematics, surveying and engineering physics at Queenwood College, a Quaker school in Hampshire. Just a year later he went to the University of Marburg, in Germany, to study mathematics, physics and chemistry, graduating in 1850. One of his professors in Marburg was Robert Bunsen, of burner fame. After a spell at the University of Berlin, he returned to England, where he was made a Fellow of the Royal Society in 1852 and became Professor of Natural Philosophy at the Royal Institution (RI) in 1853; in 1867 he succeeded Michael Faraday as Director of the RI, a post he held until he retired in 1887.

Among Tyndall's many pieces of work he explained how the blue colour of the sky is caused by the way light is scattered in the atmosphere, did pioneering investigations of germs, and wrote the first popular account of the kinetic theory of heat (*Heat Considered as a Mode of Motion*, published in 1863). His lectures in America in 1872 and 1873 were not only hugely popular but also a great financial success; Tyndall gave all the profits to the establishment of a trust fund for the benefit of American science. He was also one of the movers behind the inauguration of the science journal *Nature*. But what we are interested in here is his study of the way carbon dioxide interacts with infrared radiation – which grew out of a visit to the Swiss Alps in 1849.

This first visit to the Alps was intended primarily as a holiday, but Tyndall – like many of his contemporaries – became fascinated by glaciers, and made annual visits for several years to study these rivers of ice. At that time, there was great interest in the recent discovery that the Earth has experienced one or more great Ice Ages, when it has been much colder than it is today; with Tyndall's interest in both glaciers and heat it was natural that he should try to find an explanation for why the Earth should sometimes go into deep freeze. In the spring of 1859 he began to study the way various gases interact with infrared radiation. His big discovery was that 'perfectly colourless and invisible gases and vapours' such as nitrogen, oxygen, carbon dioxide (which he called carbonic acid) and water vapour behaved very differently when exposed to 'radiant heat'. He found that although infrared radiation passes right through oxygen, nitrogen and hydrogen with scarcely any effect, carbon dioxide, water vapour and ozone (the tri-atomic version of oxygen) all absorb infrared radiation very effectively. Water vapour is the strongest absorber of this radiant heat, and since there is a lot of water vapour in the Earth's atmosphere, Tyndall concluded that it is the most important gas in controlling the temperature at the surface of the Earth.

Tyndall discussed his ideas in a presentation to the Royal Institution in 1859, and presented his detailed results to the scientific community in a lecture given to the Royal Society in February 1861 and published in their *Philosophical Transactions*. 'Those who like myself have been taught to regard transparent gases as almost perfectly diathermanous, will probably share the astonishment with which I witnessed the foregoing effects,' he told his audience. After pointing out the powerful heat-absorbing effect of water vapour, he concluded that changes in the influence 'exercised by the aqueous vapour . . . must produce a change of climate' and that 'similar remarks would apply to the carbonic acid diffused through the air.' Furthermore, 'such changes in fact may have produced all the mutations of climate which the researches of geologists reveal.'

Tyndall elaborated and refined his argument in a series of papers on the subject in the *Philosophical Magazine* in the early 1860s – nearly 150 years ago. In his words:

> The solar heat possesses, in a far higher degree than that of lime light,[1] the power of crossing an atmosphere; but, when the heat is absorbed by the planet, it is so changed in quality that the rays emanating from the planet cannot get with the same freedom back into space. Thus the atmosphere admits the entrance of the solar heat, but checks its exit; and the result is a tendency to accumulate heat at the surface of the planet.

1. 'Limelight' was widely used in stage lighting in the nineteenth century. It is a brilliant white light produced by heating calcium oxide ('lime') in a flame of oxygen and hydrogen.

In modern language, the argument runs like this. Sunlight passes through the atmosphere almost unaffected, because it is mostly in the wavelengths of visible light, and warms the surface of the Earth. Most of the energy in sunlight is in the form of visible light because the Sun is so hot – its surface is at a temperature of about 6,000°C. Cooler objects radiate energy at longer wavelengths, and hotter objects at shorter wavelengths, in every case with a peak at a wavelength corresponding to the temperature of the object, following a rule known as the black body law. Because the surface of the Earth is much cooler than the surface of the Sun, it radiates energy at much longer wavelengths, in the infrared. And a great deal of this infrared radiation from the surface of the Earth is absorbed by gases such as water vapour and carbon dioxide in the air, warming the atmosphere. When the atmosphere in turn radiates its warmth away, some goes out into space and some goes back down to the surface of the Earth, keeping it warmer than it would otherwise be. This is the mis-titled atmospheric 'greenhouse effect' that keeps the Earth 33°C warmer than the Moon.

In a paper published in 1862, Tyndall used a rather different, but no less dramatic, analogy:

> As a dam built across a river causes a local deepening of the stream, so our atmosphere, thrown as a barrier across the terrestrial [infrared] rays, produces a local heightening of the temperature at the Earth's surface.

Significantly, though, as he spelled out elsewhere, 'the dam, however, finally overflows, and *we give to space all that we receive from the sun*.'[1] For a particular concentration of greenhouse gases in the atmosphere, the planet reaches an equilibrium at a temperature where the cooler outgoing radiation exactly balances the hotter incoming radiation; increasing the concentration of greenhouse gases in the atmosphere is equivalent to raising the height of the dam, thereby deepening the water level (increasing the temperature) at which this happens. Tyndall described water vapour as:

> A blanket more necessary to the vegetable life of England than clothing is to man. Remove for a single summer-night the aqueous vapour from the air . . . and the sun would rise upon an island held fast in the iron grip of frost.

1. Our italics.

Tyndall suggested that changes in the amount of water vapour and carbon dioxide in the atmosphere could therefore cause the kind of climate changes represented by Ice Ages ('all the mutations of climate which the researches of geologists reveal'), but made no detailed calculation of the size of the suggested effect.

The next step in the development of the understanding of global warming also came about through the search for an explanation of Ice Ages, when a Swedish chemist turned his attention to the puzzle for a bit of light relief from his main work, and came up with the first calculation of what effect either halving or doubling the amount of carbon dioxide in the air would have on the average temperature at the surface of the Earth.

Svante Arrhenius was born in 1859, the year John Tyndall began to study the way that different gases absorb radiation, and died in 1927. He came from Uppsala, where his father was an estate manager. Although he went to Uppsala University in 1876 intending to study chemistry, he found the teaching in the chemistry department there so bad that he switched to physics for his first degree, then moved on in 1881 to Stockholm to work for a PhD in physical chemistry. This educational path gave him a thorough grounding in physics as well as chemistry, which he later put to good use in his work on global warming, even though he regarded this merely as a hobby. Arrhenius received his doctorate in 1883, but his thesis did not receive top marks, and this made it difficult for him to get a permanent academic post. For five years he travelled around Europe on a scholarship from the Swedish Academy of Sciences, again broadening his experience more than a conventional career path might have, and by the end of the 1880s he was recognized as one of the leading chemists in the world. He returned to Stockholm in 1891 as a lecturer at the Technical Institute (Högskola), a forerunner of Stockholm University, where he became a Professor in 1895. Arrhenius received the Nobel Prize for chemistry in 1903, and in 1905 he became Director of the Nobel Institute, where he stayed until shortly before he died. From the time he returned to Stockholm, alongside his work in chemistry Arrhenius developed interests in astrophysics, the origin of life (he suggested that the Earth might have been 'seeded' with spores from space, an idea known as *panspermia*) and climate change.

Climate change was something of a hot topic at the end of the nineteenth century because of the realization, as we have mentioned, that the Earth has experienced several glaciations in the relatively recent geological past. Although other people had noticed the evidence of extensive glaciation before, the person who really started the study of Ice Ages rolling was Louis Agassiz (1807–1873) who, as the thirty-year-old President of the Swiss Society of Natural Science, astonished the audience at his Presidential Address in 1837 by launching into an impassioned lecture on Ice Ages – indeed, it was in this lecture that the term 'Ice Age' was introduced. His colleagues took some convincing, but Agassiz went out on the campaign trail with enthusiasm, typified by this extract from his book *Étude sur les glaciers*, published in 1840:

> The development of these huge ice sheets must have led to the destruction of all organic life at the Earth's surface. The ground of Europe, previously covered with tropical vegetation and inhabited by herds of great elephants, enormous hippopotami, and gigantic carnivores, became suddenly buried under a vast expanse of ice covering plains, lakes, seas and plateaus alike. The silence of death followed . . . springs dried up, streams ceased to flow, and sunrays rising over that frozen shore . . . were met only by the whistling of northern winds and the rumbling of the crevasses as they opened across the surface of that huge ocean of ice.

No wonder many people in the later nineteenth century worried about the return of the ice!

Geologists slowly gathered evidence that the Earth has experienced not one but several glaciations in the past few million years, and that there were indications of a repeating rhythm of Ice Ages and warmer intervals now known as Interglacials. Today, it is clear that more extensive ice cover than we see on Earth now has been normal for at least the past five million years, and warmer Interglacials like the one we live in have been relatively short-lived departures from the long-term average. But those discoveries lay far in the future when nineteenth-century scientists struggled to find an explanation for the advance and retreat of the ice.

One idea was that the climate of the Earth is affected by changes in the balance of the seasons caused by the way the Earth wobbles (like a wobbling spinning top) as it orbits around the Sun, and by small changes in the orbit itself. The orbit changes from more circular to slightly more elliptical, because of the gravitational influence of the other planets. The overall effect is that even though the average amount of heat received by the entire Earth from the Sun in the course of each year stays the same, sometimes our planet experiences hot summers and very cold winters for thousands of years in succession, while at other times in the cycle there are millennia with less difference between the seasons. The timescale of these changes is just the same as that of the rhythm of Ice Ages and Interglacials.

The first person to suggest a link between these astronomical rhythms and Ice Ages was a French mathematician, Joseph Adhémar (1797–1862), who presented it among a hotchpotch of rather confused ideas in a book, *Révolutions de la mer*, published in 1842. But the person who really put the astronomical theory of Ice Ages on the scientific map was a Scot, James Croll, born in 1821. Croll is a fascinating figure who came from a poor crofting family and had no formal education, but who worked his way up to obtain a post with the Geological Survey of Scotland in 1867 and be elected as a Fellow of the Royal Society in 1876. His first scientific paper on Ice Ages was published in 1864, but the clearest presentation of his idea appeared in a book, *Climate and Time*, published in 1875.

In a nutshell, Croll argued that what you need to start an Ice Age is a sequence of very cold winters, allowing snow to pile up at high latitudes and be compressed into ice sheets by the addition of more snow each year. According to his model, if you put the calculations of the astronomical rhythms in you discover that the Northern Hemisphere should have been warming out of an Ice Age between about 100,000 and 80,000 years ago, when its winters were relatively mild and summers were relatively cool. But by the end of the nineteenth century geologists had evidence that at exactly that time the temperature was declining – the world was plunging into the latest Ice Age.

That should have been a clue – it is now clear that for millions of years it has always been cold enough for snow to fall in winter with the potential to build ice sheets. The *natural* state of the planet has been in an Ice Age, so it has only warmed into an Interglacial state when summers have been really hot – hot enough to melt the ice.

Cooling northern summers between 100,000 and 80,000 years ago allowed normal service to be resumed after a short-lived warm spell. But that connection, which we discuss in more detail later, was not made in the nineteenth century, and Croll's astronomical model for Ice Ages fell from favour following his death in 1890, before Arrhenius turned his attention to the puzzle of Ice Ages.

Arrhenius had two big advantages over pioneers like Tyndall when he started to investigate how heat is trapped by the atmosphere of the Earth. By that time, Josef Stefan (1835–1893) had discovered the mathematical law which relates the temperature of an object to the amount of energy it radiates, and there were accurate measurements of how much energy is trapped in this way by different gases, thanks to Samuel Pierpoint Langley, who invented the bolometer – an instrument sensitive enough to measure temperature differences of one hundred-thousandth of a degree Celsius when warmed by radiation over a wide range of wavelengths. Developments of these bolometers are flown on many satellites today to monitor from space the changing heat balance of the Earth (and, indeed, on missions to Mars and other planets).

The idea of launching his instruments into space would surely have appealed to Langley, who among other things was an aeronautical pioneer, the man after whom NASA's Langley Research Center and the Langley Air Force Base are named. He was born in 1834, in Boston, Massachusetts, and was another of the nineteenth century's 'self-made' scientists. After leaving high school, he devoured science books from libraries and worked as a civil engineer in Chicago and St Louis before getting his first scientific job as an assistant at the Harvard Observatory. After a spell teaching mathematics at the US Naval Academy in Annapolis, in 1867 he settled as Professor of Physics and Astronomy at the Allegheny Observatory in Pennsylvania, where he stayed for twenty years. He then became Director of the Smithsonian Institution in Washington DC, where he spent the rest of his career; he died in 1906.

From the mid 1880s onwards, Langley carried out a series of experiments with flying machines. He is thought to have been the first person to build heavier-than-air machines, powered by steam, which were capable of sustained flight, although they were unmanned and uncontrolled. At the beginning of the twentieth century, when he was in his late sixties, Langley built an aircraft powered by a gasoline engine that was intended to carry a human pilot. There is every likelihood that it would have flown, but it was let down (literally) by its catapult launching system. Two test flights, wisely attempted over water, ended in the Potomac River; the second of these launch failures occurred on 8 December 1903, just nine days before the first successful flight of the Wright brothers' machine.

Langley's measurements of infrared absorption by the atmosphere were made in the second half of the 1880s. The bolometer he used was so sensitive that it could measure the amount of heat falling on it at different wavelengths as it moved along the spectrum – or rather, as the spread-out spectrum from a source like the Sun is moved across the bolometer. Each kind of molecule in the air radiates energy (when it is hot) or absorbs energy (when it is cooler than the radiation passing through it) over specific ranges of wavelengths, known as emission or absorption bands; the bands are the same whether the molecules are emitting or absorbing. Spectroscopic observations revealed the presence of many absorption bands associated with both water vapour and carbon dioxide, and by making observations of the spectra of both the Sun and the Moon, Langley was able to pinpoint how much radiation gets trapped in the air. The question Arrhenius set out to answer in the second half of the 1890s was, how does this trapped energy affect the temperature at the surface of the Earth?

This work involved a great deal of tedious calculation in those days before the advent of electronic calculators (let alone computers), but as early as 1895, in a paper presented to the Stockholm Physical Society, Arrhenius was able to write that:

> Temperature of the Arctic regions would rise about 8 degrees or 9 degrees Celsius, if the carbonic acid increased to 2.5 to 3 times its present value. In order to get the temperature of the ice age between the 40th and 50th parallels, the carbonic acid in the air should sink to 0.62 to 0.55 of present value (lowering the temperature 4 degrees to 5 degrees Celsius).

At that time, Arrhenius was chiefly interested in finding an explanation for Ice Ages – if you like, global cooling. He developed his ideas and carried out further extremely tedious calculations involving a lot of work over the next few years, publishing a weighty textbook, which received little attention, in 1903. By that time, he was beginning to appreciate the importance of global warming, and realized that human activities had the potential to change the climate. His colleague Nils Ekholm pointed out in 1899 that human activities had the potential to double the amount of carbon dioxide in the atmosphere, and that this would 'undoubtedly cause a very obvious rise of the mean temperature of the Earth'. Arrhenius picked up on the idea, and in 1904 he wrote that 'the slight percentage of carbonic acid in the atmosphere may, by the advance of industry, be changed to a noticeable degree in the course of a few centuries.' But from his historical and geographical position, this seemed like a good thing.

Arrhenius had no idea how rapidly the burning of fossil fuel would increase and how swiftly carbon dioxide would build up in the atmosphere as a result. And from the perspective of Sweden, a slightly warmer world looked desirable. In his popular book *Worlds in the Making* (1906),[1] drawing on work he had done over the previous ten years, Arrhenius explained just how cold the world would be without the 'hothouse' effect. Crucially – and significantly in terms of the development of the kind of ideas important for Gaia theory – he included feedback in his calculations. First, Arrhenius calculated that taking all of the carbon dioxide out of the air would cause the average temperature at the surface of the Earth to drop by 21°C. But he realized that such a cooling would also reduce the amount of water vapour in the air, and he calculated that this reduction in water vapour would cause a further cooling of 10°C, giving a total cooling of 31 degrees, almost all the way down to the figure appropriate for the airless Moon and close to the figure we get from modern calculations.

This is an example of a positive feedback. Cooling the Earth by whatever means takes water vapour, one of the greenhouse gases, out of the air as it condenses, and makes the planet cooler still. The same thing happens in reverse – when the world warms, for whatever reason, water evaporates from the oceans and infrared heat trapped by the extra water vapour in the air makes the warming bigger than it would be if there were no feedback. There are also examples of

1. The English translation did not appear until 1908.

negative feedbacks, which act to reduce changes and maintain the status quo; we shall come across these later.

Arrhenius was satisfied that reducing the amount of carbon dioxide in the air could easily explain why Ice Ages happened, even though that raised the question of why the amount of carbon dioxide in the air should vary. As he developed his ideas, he also pointed out that if the amount of 'carbonic acid' in the atmosphere increased slightly from its present level, this might prevent the onset of another Ice Age, make the climate of Europe more equable, stimulate the productivity of plants which need carbon dioxide in order to grow, and provide more food for the world's increasing population. 'We would then have some right,' he suggested from his chilly northern home, 'to indulge in the pleasant belief that our descendants, albeit after many generations, might live under a milder sky and in less barren surroundings than is our lot at present.'

This optimistic scenario was based on the assumption that the amount of carbon dioxide added to the air by human activities would be small compared with the natural 'reservoir' of the gas in the atmosphere. Arrhenius was able to make the comparison between anthropogenic emission of carbon dioxide and natural processes because one of his colleagues in Stockholm, Arvid Högbom, had been studying the way carbon dioxide cycles in the Earth System work, with the gas being released by volcanoes, absorbed by the oceans, and so on. Arrhenius calculated in 1896 that doubling the quantity of carbon dioxide in the air would warm the world, allowing for feedbacks, by 5°C – which, partly by luck, is very close to the best modern estimates. But he never dreamed how rapidly the world would change in the twentieth century, and thought that at the rate human activities were releasing carbon dioxide in the 1890s it would take three thousand years for industrial activities to produce such a rise. By 1906 he had already revised this estimate down to a few centuries; in fact, the amount of carbon dioxide in the air has already risen, since Arrhenius' day, by more than 25 per cent, and the doubling is projected to take place before the end of the present century.

Hardly anyone, though, believed Arrhenius' claim that adding carbon dioxide to the air would make the world warmer. The accepted wisdom of the time – which we now know was based on inadequate measurements made with spectrographs less sensitive than those of today – was that in the parts of the infrared spectrum where carbon dioxide absorbs energy the bands were already 'saturated', with all the radiation being absorbed, so there was none left over to be absorbed by additional carbon dioxide. There was also an assumption that any carbon dioxide added to the air by human activities would be absorbed by the oceans. This is true up to a point, but it takes the oceans thousands of years to absorb what human activities can now release in a century.

While the idea of carbon dioxide as a regulator of climate and possible cause of Ice Ages languished after the work of Arrhenius, the rival astronomical theory suffered a similar modest rise and rapid fall from grace. The idea was revived by the Serbian Milutin Milankovitch (1879–1958), who devoted most of his career to refining the calculations of how the astronomical effects we have described change the amount of heat arriving at the Earth from the Sun (the *insolation*) at different latitudes and in different seasons. By the time the First World War broke out, Milankovitch was a Professor at the University of Belgrade and already deep into his laborious calculations that would eventually reveal how insolation had changed over the past 600,000 years. In the wrong place at the wrong time, on a visit to Hungary when the war broke out, Milankovitch was interned by the Hungarians and spent four years in Budapest with nothing to do but continue his calculations. First, he came up with a mathematical model, with every number worked out by hand, describing the climate of the Earth today, then he adapted it to cover Venus and Mars as well.

The fruits of all these labours appeared in a book published in French in 1920. At first, the only person who appreciated the value of what Milankovitch had achieved was a Russian-born German meteorologist, the 76-year-old Wladimir Köppen, who wrote to Milankovitch from Hamburg, initiating a long correspondence. It was Köppen who provided Milankovitch with the key insight that the way to start an Ice Age in the Northern Hemisphere is to have cool summers, rather than very severe winters. Armed with this insight, Milankovitch embarked on another calculating epic, and found that the timing matched the geological record. Over the past six hundred millennia, when Northern Hemisphere summers have been at their coolest the Alpine glaciers have advanced.

What became known as the Milankovitch Model of Ice Ages gained some currency in the 1920s and 1930s, and was summed up in a book that Milankovitch saw published in 1941 – ironically, in the German language, just at the time Yugoslavia was being invaded by German forces in the Second World War. But in truth the geological chronology was too poor to make the match between Ice Ages and insolation variations convincing to sceptics, and even if the timing of Ice Ages and astronomical rhythms matched, the size of the astronomical effect seemed too small to account for the size of the climatic fluctuations. The astronomical rhythms might be, as they were later called, the 'pacemaker' of Ice Ages, but they could not on their own *drive* Ice Age/Interglacial fluctuations. So the Milankovitch Model began to lose what modest support it had almost exactly at the time Milankovitch published his big book.

Meanwhile, the carbon dioxide model of Ice Ages had re-entered the arena of scientific debate, even if it had won few converts. One person who was convinced was the American physicist E. O. Hulburt, who pointed out the flaws in the argument about saturation of the infrared bands in a paper published in the *Physical Review* as early as 1931.[1] But meteorologists didn't read the *Physical Review*, and his paper had no impact at the time. The person who did make meteorologists at least sit up and debate the issue was a British physicist, Guy Stewart Callendar, who had learned thermodynamics (the science of heat) almost literally at his father's knee and turned his attention to the carbon dioxide greenhouse effect in the late 1930s.

Callendar was born in 1897, and died in 1964. His father was Hugh Callendar, at the time of Guy's birth Professor of Physics at McGill College in Montreal, but soon to return to his native England as Professor of Physics first at University College London, then at the Royal College of Science (now Imperial College), also in London, where he was head of the Physics Department from 1908 to 1929.

1. The earlier argument against the idea that adding carbon dioxide to the air would make the globe warmer was also wrong because it failed to take account of the way absorption of infrared radiation is affected by temperature, and the temperature of the atmosphere changes considerably with altitude.

The elder Callendar was an authority on thermodynamics and steam power, especially the application of thermodynamics to steam turbines – hugely important to industry and shipping. He had been elected as a Fellow of the Royal Society in 1894. Among his many other scientific interests, Hugh Callendar also invented a kind of thermometer based on monitoring the way the resistance of a platinum wire changes with temperature; this formed the basis of the kind of chart recorders used ever since to record changes in temperature continuously on long rolls of paper.

Under the influence of his father, Guy Callendar studied engineering in London then worked as one of Hugh Callendar's research assistants from 1923 to the end of the 1920s. After his father died in 1930, the younger Callendar took over some of his lecturing duties and carried out research on steam turbines and fuel cells. From 1942 until he retired in 1957 he worked for the Ministry of Supply. But all the time his real scientific passion was the study of weather and climate, which he carried out in his own time, strictly speaking as an amateur meteorologist, but one with a thorough grounding in physics and in particular in thermodynamics. His biographer James Fleming sums Callendar up as 'a well-trained, extremely competent, pensive, and somewhat reclusive engineer, a loving husband and devoted father'.

Callendar's 'hobby' bore fruit in 1938, when he was already in his forties, when he presented a paper to a meeting of the Royal Meteorological Society. He had been collecting weather statistics for years – in particular, temperature data from around the globe, going back to the end of the nineteenth century. Other people had found hints in the partial records available to them that the world had warmed during the first third of the twentieth century, but the statistics Callendar had gathered – from some 200 weather stations around the world – established this beyond any doubt. That would have been enough of an achievement for most amateur meteorologists; but Callendar went further. He told the meeting that he had an explanation for the warming – the addition of carbon dioxide from human activities into the atmosphere of the Earth was adding to the greenhouse effect. He pointed out that over the previous fifty years burning fuel (mostly coal, in those days) had released about 150 billion tons of carbon dioxide into the atmosphere, and that three quarters of it was still there, representing an increase in the concentration of carbon dioxide in the air of 6 per cent between 1900 and 1936.

When he had first come across the work of Arrhenius, Ekholm and other early proponents of the idea of human-induced global warming, Callendar had not been persuaded by their arguments. But he knew that much better measurements of the properties of the so-called greenhouse gases had been obtained since the beginning of the twentieth century, so rather than dismiss the notion out of hand, in the best scientific tradition he carried out his own calculations to test the idea. Contrary to his expectations, he found that the effect was real. According to Callendar's calculations of the greenhouse effect, using the latest data on infrared absorption and information about the structure of the atmosphere, his 1938 estimate was that doubling the amount of carbon dioxide in the air would produce a rise in global mean temperature of at least 2°C, although he noted that the effect might be 'considerably greater'. Since the world had warmed by just one sixth of a degree since 1900, this meant that according to his calculation the anthropogenic greenhouse effect could account for between two thirds and three quarters of the warming. This is an important point which was appreciated even by this early pioneer of global warming studies – nobody claims that *all* changes in climate, even today, are a result of human activities. There are natural fluctuations as well, just as there have always been, with the human influence superimposed on them. What Callendar was saying was that the warming of the world between 1900 and 1936 was three times greater than it would have been without human interference.

Callendar's words are eerily similar to those being used by climatologists today, some seventy years later:

> If any substance is added to the atmosphere which delays the transfer of low-temperature radiation, without interfering with the arrival or distribution of the heat supply, some rise of temperature appears to be inevitable in those parts which are farthest from outer space.

But how should we react to the prospect of such a rise in temperature? Here, Callendar, like Arrhenius, struck a very different note from his modern counterparts. In the late 1930s, global warming still seemed like a good thing, and in any case Callendar estimated that the average global temperature increase caused by human activities would only be about one degree over the next two hundred years. So:

> The combustion of fossil fuel, whether it be peat from the surface or oil from ten thousand feet below, is likely to prove beneficial to mankind in several ways, besides the provision of heat and power . . . the return of the deadly glaciers should be delayed indefinitely.

In 1939, Callendar reported that 'The five years 1934–38 are easily the warmest such period at several stations whose records commenced up to 180 years ago.' But this was still seen as very much a good thing, whatever the cause of the warming.

Callendar continued to study the role of carbon dioxide and other greenhouse gases throughout his life. In 1941 he published a paper reviewing the spectroscopic measurements and drawing attention to the absorption bands of carbon dioxide itself, water vapour, nitrous oxide and ozone, all components of the Earth's atmosphere. The overall effect of his work was to make professional meteorologists appreciate that the absorption of infrared radiation by carbon dioxide in the atmosphere really is important, and a problem worth studying. For this reason, some climatologists have tried to promote the use of the term 'Callendar effect' for what is usually known as the anthropogenic greenhouse effect; but they are fighting a lost cause, since, as we have acknowledged, the term 'greenhouse effect' is just too catchy. In Callendar's lifetime, though, the study of this greenhouse effect was still only a minority interest. The idea of a human-induced global warming seemed far-fetched, and those who thought it was real felt it was probably a good thing – not least because the Northern Hemisphere cooled (for reasons we shall discuss later) in the three decades following Callendar's 1938 presentation at the Royal Meteorological Society. Callendar himself continued revising his work and publishing papers on the greenhouse effect until his death in 1964, although nothing had the impact of his early papers. But in 1941, the year Callendar published his review of the data on carbon dioxide absorption and Milankovitch published his epic book on insolation cycles, a young man who would eventually put all of these ideas into their proper global perspective was just finishing his degree at the University of Manchester.

2

A Child of His Time

James Ephraim Lovelock was born in the afternoon of 26 July 1919 – a product, he believes, of the celebration of Armistice Night at the end of hostilities in the First World War on 11 November 1918. He was very much a child of his time. He came from a working-class background, but because so many men were away in the trenches his mother, Nell, an intelligent woman who would otherwise have had no opportunity to use her abilities, had a responsible job with Middlesex County Council during the war years. This helped to fuel her ambitions for her son, who, like many of his contemporaries, was brought up to believe that education offered a way out of working-class drudgery. Nell had actually won a scholarship to Islington Grammar School, but had had to turn it down in order to start work at thirteen; the job she was forced into was putting labels on pickle jars. She grew up to become an early feminist, proud of her independence, who like many women of her generation smoked cigarettes 'almost as a badge of honour'.[1] But she always regretted her missed educational opportunities; the family was determined that Jim would never have to suffer such a blow. England may not have been the land fit for heroes that returning soldiers were promised, but in the 1920s and 1930s the old barriers to upward mobility were at least beginning to break down.

Jim's father, Tom, was 47 when Jim was born and had been too old to fight in the war. Long before Jim arrived, Tom had served six months hard labour for poaching – in Reading Gaol, the same place

1. Unless otherwise indicated, quotations are from our interviews with James Lovelock or from his book *Homage to Gaia*.

that Oscar Wilde was incarcerated. But the details were never discussed. Tom had a job at the South Metropolitan gasworks in Vauxhall, and had been married before. His first wife suffered from depression, and in the relatively unenlightened first decade of the twentieth century had been confined in a lunatic asylum, where she died in 1914. By then, Tom and Nell, who was fourteen years younger than Tom, had already met and fallen in love, though the relationship had gone no further than holding hands. Able to marry at last in 1914, they had an idyllically happy few years; Tom loved art, Nell was passionate about classical music, and both of these were in good supply in London in spite of the war. They lived in a flat in Clapham, but Jim was born in the house of his maternal grandparents, Alice and Ephraim March, in Letchworth. Although the times were changing in England after the First World War, as this indicates the extended family was still much more important than it is today, and 'Nana March' would play a big part in Jim's early life.

The name Lovelock, thought to derive from the Anglo-Saxon 'lovelich', meaning 'affectionate', is not uncommon, and Jim is proud of the connection with a distant relative, John Edward ('Jack') Lovelock, who was the first New Zealander to win an Olympic Gold Medal, setting a world record time in the final of the 1500 metres in Berlin in 1936. Jack's near-namesake was a sixteen-year-old schoolboy by then, only nine years younger than the other J. E. Lovelock. But his life had already been less ordinary than most.

With Nell having lost her job with the return of men from the war, Jim's parents, fired with optimism, opened a shop selling paintings and other objets d'art in rented premises on Brixton Hill. The enterprise always struggled to make a profit in a suburb that had declined from its former gentility as London had expanded, and Tom Lovelock continued to work for the South Metropolitan, now as a collector walking from house to house to empty coins from meters. With Nell fully occupied with the shop, little Jim spent the first six years of his life in the care of Nana March, 'a small plump cockney woman endowed with a surfeit of love'. Lovelock's grandfather Ephraim worked as a bookbinder; he came from working-class stock near Dagenham, in Essex, but by the time Jim came on the scene he had prospered sufficiently through his skill at his craft to own a four-bedroomed house in the 'garden city' of Letchworth, some 30 miles north of London. Apart from Nell, their oldest child, Alice and Ephraim had three other married daughters and a grown up son who lived in London. One of their daughters, Jim's aunt Kit, was married to Hugo Leakey, a cousin of Louis Leakey, who later made dramatic discoveries of early human remains at Olduvai Gorge in Tanzania.

Jimmy, as she called him, was Nana March's first grandchild, and he had her undivided attention and love for nearly six years. He remembers these as years 'full of happiness and sunshine', spent largely running free or in the company of adults rather than children, growing up secure and self-sufficient, with an inquisitive nature and vivid imagination. He was briefly enrolled in a private school, but the experiment didn't last long. After the teacher showed her class the various poisonous plants that grew on the local common, intending to warn them of the dangers, Jimmy, curious to learn more about their effects, got hold of some deadly nightshade berries and tried to persuade some girls from another class to eat them. Fortunately, the experiment was interrupted by the arrival of another teacher, and the miscreant was sent home in disgrace. On another occasion, he was caught begging for pennies outside a sweet shop, using the highly successful ploy of claiming that his father was out of work. But what he now recalls as the most important event from his time in Letchworth came at Christmas in 1923, when his father gave Jim a box of electrical bits and pieces – a torch bulb, wires, batteries, electric bell and so on. The experiments that he carried out with that box of bits, and the questions the experiments raised that none of the adults he knew could answer, were, he is convinced, what led Lovelock into a life of science. Although not yet five years old, 'I realized that I would have to find the answers myself.'

What seems to have been an idyllic childhood in Letchworth came to an end in 1925, when Lovelock's grandfather retired. As he had no adequate pension, he had to sell the house, and with the money invested to provide for his retirement, Alice and Ephraim moved in to rooms above the shop in Brixton. 'The living conditions,' says Lovelock, 'were primitive.' On the ground floor, behind the shop, there was a living room and his parents' bedroom. Jimmy's bedroom, like his grandparents' rooms, was upstairs; but there were no washing or cooking facilities above the ground floor. Leading off from the living room was a scullery, where most meals were cooked on a gas ring, and beyond that a lavatory with a hand basin for washing. Outside, in the paved yard, there was a draughty wooden shed which contained a gas oven (used only to cook Sunday lunch) and a galvanized iron hipbath, used in turn by all the family. Especially in winter, bathing was a rare event for the adults; but Jim's mother gave him a daily bath in front of the gas fire in the living room, in a forlorn attempt to remove the grime that accumulated everywhere from the coal fires of London.

This grime, which was an inevitable consequence of city life in the 1920s, had a far deeper significance, we now know, than the inconvenience of the need for daily baths in front of the fire. Although London was the most notorious example, the pall of pollution from the coal fires not just of domestic hearths but of industry and power stations[1] spread across northwest Europe, and much of the Northern Hemisphere. At ground level, this was responsible for the 'pea soup' fogs of Victorian London, later dubbed 'smog' from a combination of the words 'smoke' and 'fog'. In the worst of these smogs, it was sometimes literally impossible to see a hand held at arm's length in front of the face, and thousands died from illnesses such as bronchitis. It didn't help that bedrooms were draughty and unheated. In winter, it was common to wake up to find the window covered on the inside with a delicate tracery of frost, from the moisture in the air breathed out by the sleeper; it wasn't uncommon for water left in a glass by the bedside to freeze over. Both young Jim and his mother (a cigarette smoker) were severely affected by the smogs and the difficulty of keeping warm in winter; but Lovelock doesn't remember his father ever being ill.

More insidiously, the spread of high-altitude pollution in the form of fine particles drifting around the Northern Hemisphere reflected away some of the incoming warmth of the Sun and cooled the land below – an effect now sometimes referred to as global dimming, although we prefer the term solar dimming. This is one of the principal

1. The gas for the fire that kept Jim warm while he bathed also came from coal, processed at the local gas works by the company that employed his father.

reasons why the world failed to warm in line with the predictions of Arrhenius and Callendar in the middle decades of the twentieth century. For a time, the particulate pollution caused by industry counterbalanced the enhanced greenhouse effect caused by industry. Before long, the balance would tip in favour of global warming. All of this would have a profound effect on the life and career of young Jim, blissfully ignorant of what was going on in the atmosphere of the Earth as he splashed about in his bath in front of the fire.

To modern eyes, the living conditions of this extended family may indeed seem primitive; in fact, by the standards of the day they were reasonably well off at first, as the shop was, for a time, proving modestly successful. The Lovelocks even employed an assistant to help in the shop, where the work ranged from framing photographs to selling pictures and prints. But relative financial comfort came only as a result of hard work. In addition to his day job, in the evenings Tom Lovelock restored paintings in a basement beneath the shop. Although it would be an exaggeration to say that Tom and Nell played as hard as they worked, while the times were good Nell frequently attended concerts, and the couple were able to take a holiday in Europe each year, usually two weeks on a Thomas Cook tour. They visited Rome, Madrid, Paris and the Swiss lakes – but their son always stayed behind with his grandparents. It never would have occurred to them to take a child on such a trip, and Jim never felt hard done by. His own holiday memories are of excursions with his grandparents, the most memorable being trips down the River Thames from Westminster Pier to Margate on the paddle steamer *Royal Sovereign*. He was fascinated by machinery, including steam-hauled trains, and loved to watch the huge pistons of the ship's engine at work. The river was crowded with shipping in the 1920s, both cargo vessels and the great passenger liners that docked at Tilbury, and there was plenty to see on the banks of the river. This was just as well, since at Margate the day trippers had only an hour to fill before it was time for the journey home.

By this time, Jim had started to attend a private primary school. Initially, this was as much of a disaster as his first encounter with school back in Letchworth, but for a different reason. His first class teacher, Miss Tierney, made such an impression that eighty years later he refers to her with feeling as 'an embittered Irishwoman'. She was the kind of teacher who liked uniformity and hated a precocious child who had the temerity to have his own ideas and was not afraid to speak up for them. The result was frequent caning across Jim's hands and fingers, until he simply decided to stop going to school and set off each morning instead to play around Brixton Hill. Of course, this truancy was soon noticed, but the school had enough sense to see the root of the problem and moved Jim into a class where the teacher, a Miss Plumridge, was 'a plump motherly woman who referred to me always as "The brand plucked from the burning".' Lovelock still seems rather proud of the description, and says that although discipline was still important in the classroom, it was now discipline with a purpose, and he remembers the school as being 'wholly wonderful'. He was encouraged to make rapid progress, soon learning to read fluently, and became an early fan of science fiction, especially the books of Jules Verne, Olaf Stapledon and H. G. Wells, which he borrowed from the local library. Verne wrote 'good adventure stories', but not 'real science fiction'. To Jim, it was H. G. Wells who invented real science fiction, and he particularly remembers the impression made by *The Time Machine*, with its apocalyptic vision of a scorched future Earth – almost a scientific prediction of the actual fate of the Earth when the Sun swells to become a Red Giant star, and the warming overwhelms the mechanisms of Gaia. 'That really stayed in my mind.'

By the end of his time at the school, Jim had acquired an early fluency with both words and numbers that would stand him in good stead throughout his life; he is firmly of the opinion that such literacy and numeracy have to be instilled into a child at an early age, so that they become as natural a process as walking or riding a bicycle. Then, they can provide the bedrock on which to build a good education; but without early literacy and numeracy no later efforts will ever be quite as effective.

All of this early educational progress was, of course, encouraged by Jim's family, especially his mother, who still regretted her own missed educational opportunities and pressed for Jim to move on to the local grammar school at the earliest possible age. Lovelock did not enjoy the experience, and still sees it as a mistake. It was natural for working-class people in the 1920s to believe that a 'good education' for their children would provide opportunities for them to rise in society, and to some extent this was true. But Lovelock disparages the notion that a good education can turn any child into a genius (as he puts it, 'any girl into a Jane Austen or any boy into a Charles Darwin'). 'You have to have some raw material to work with,' he says. At the age of nine, Jim was extremely bright, but he was too immature to benefit from being thrust from the comfortable environment of a primary school where he was doing well into a grammar school where he was the youngest and most insignificant boy and expected to work within a system that offered little scope for individuality.

By this time, Jim's mother had also introduced him to another, very different, 'system' which encouraged his individuality. Her wartime duties as secretary to the Clerk of Middlesex County Council had included attending hearings of conscientious objectors, where she noticed that the only ones shown any respect were the Quakers. Although both Nell and Tom were 'staunchly agnostic' she became interested in the Quaker philosophy and enrolled Jim at the Society of Friends Sunday School in Brixton. This was nothing like the Sunday Schools of other religions, and Lovelock still recalls with pleasure the stories, discussions and other entertainment that took place on Sunday afternoons – including Felix the Cat movies and other cartoons. In time, his mother became a Friend herself, although as he grew older Jim moved into 'a comfortable agnosticism as science began to fill the empty files of my mind'. But there was nothing comfortable about his introduction to secondary school.

Lovelock hated the grammar school. It was a good school of its kind, but it just didn't suit him (although we suspect his experience might have been different if his mother had waited another couple of years before rushing him through its gates). He describes himself as having been a weedy child who suffered badly in the filthy air of London in winter, when he looked forward to bouts of bronchitis and even pneumonia because they freed him from the tyranny of the school and gave him time at home with his beloved books. When he did attend school, he was wary at first of being a potential target for bullies, but says that it turned out that bullying was virtually non-existent. The boys were all selected by ability, and all of them respected intelligence and academic ability. Jim hung out with a crowd of other sickly specimens, among whom he had some status as their own 'mad scientist'. Although by nature a loner, Lovelock soon realized the value of social engagement with his peer group, and claims that most of the scientific knowledge he gained at the time came from books and interaction with his peers. The grammar schools of the day were only open to an elite group of boys who passed a tough entrance exam, and in his early teenage years Lovelock would discuss the hot scientific topics of the day (the early 1930s) with his fellow pupils – developments such as the particle accelerator used by the British scientist John Cockcroft and his Irish colleague Ernest Walton to 'split the atom'. But mostly Lovelock regarded school as a place where he had to mark time until he was old enough to get a job. Even Nell's dreams did not extend as far as imagining her son could go to university.

Jim could imagine the prospect of going to university, but the thought appalled him. By the age of twelve, he was determined to become a scientist, but saw the prospect of another six years at the school he hated, followed by more years at university, as intolerable. He would do things his own way, because he knew what was right for him. At this early age, Lovelock decided that although there was a legal requirement for him to attend school from Monday to Friday, the rest of his time was his own. He refused to do homework or to attend 'compulsory' sports on Saturdays, in effect working to rule like a militant trade union member, and as a result for the next two years he was repeatedly punished, by both caning and being made to write out hundreds of lines. The fact that he did well in exams in spite of refusing to do homework only annoyed his teachers more; but by the time he was fourteen they had realized that no punishment they could inflict would make him change his ways, and they essentially left him to his own devices.

There were a very few teachers who stood out from the rest. Lovelock remembers with affection one inspiring teacher of French, and in particular Harold Toms, the chemistry teacher, who was the only master in the school with a PhD. But these were isolated exceptions.

During the time Lovelock was painfully reaching his own unique accommodation with the school, the family's circumstances changed.

At the beginning of the 1930s, the economic depression hit the shop hard, and even after the shop assistant was 'let go' it lost money steadily; but the Lovelocks couldn't just walk away from it, as they had a long lease on the premises. They were fortunate to find an art lover who took over the business from them in 1932, leaving them with just enough money to purchase a small house in Orpington, on the Kent borders about 10 miles to the southeast of central London as the crow flies. About the same time, on his own initiative Jim made a significant contribution to the family finances. Against the advice of his schoolmaster, he applied for a County Scholarship which would cover the cost of his school fees. With the help of the school secretary, he filled in an application form and duly sat the necessary examination, which he passed. The response of the school, rather than praising him for his initiative, was to place him in the lowest of the three academic streams at the school. Apart from being a slightly healthier environment away from the smogs of Brixton, the move to Orpington made little difference to Jim, who stayed on at the same school but now had a daily journey of over an hour each way by train and on foot. So he made no friends in Orpington, and never thought of it as his home town, continuing to plough his own furrow.

Lovelock's own devices actually involved far more work than the homework he refused to do, but he was studying what he wanted and when he wanted, mostly using books borrowed from the library. In the 1930s, and for decades afterwards, public libraries were still a treasured resource containing books such as Jeans' *Astronomy and Cosmology*, Soddy's *The Interpretation of Radium*, and Wade's *Organic Chemistry* (a key influence on the young Lovelock) as well as the science fiction he had been reading since he was eight. It was after he had exhausted the science fiction, 'reading at an enormous rate', that Jim 'wandered down into the basement where the real science was kept'. The books at home in Orpington, which he also devoured, tended to be equally serious, dominated by the yellow jackets of the Left Book Club published by Gollancz and the works of George Bernard Shaw; Nana March was less politically inclined and enjoyed romantic fiction, but no teenage boy would bother with that.

Lovelock's only regret about his unusual relationship with school is that it left him without a thorough grounding in mathematics or any foreign languages; but he managed well enough without them. He has bigger regrets about the way the modern culture of safety first has made it impossible for young people to handle many of the chemicals and carry out many of the experiments that helped to fire his enthusiasm for science in general and chemistry in particular. Substances that he could obtain legally and experiment with for himself as a young teenager are now only seen by schoolchildren, if at all, when demonstrations are carried out by teachers. Lovelock delights in pointing out that according to statistics gathered by the Royal Society of Chemistry chemists actually have longer than average lifespans. There is an obvious possible reason for this. By coming into contact with dangerous substances and learning how to handle them, chemists become more adept at risk assessment in everyday life; by contrast, if children are never exposed to mildly dangerous activities, not even allowed to walk to school, they never learn how to assess risk and can be in big trouble when they eventually venture out into the world alone.

It wasn't just chemistry that interested the teenage Jim Lovelock. Although he found the physics taught at school excruciatingly dull, at home, as well as reading books by the likes of James Jeans and Frederick Soddy, he built his own short-wave radio receiver from the instructions in a hobby book given to him as a Christmas present. He made many of the components himself, winding coils of wire on jam jars or pencils, and was delighted to be able to listen to radio stations in America, Russia and other countries. This handmade radio set would be the forerunner of the many sensitive scientific instruments that Lovelock later invented and built himself. 'It was an important part of scientists' growing up in those days; you had much more opportunity as a child to get the feel of "hands on". All of that helped enormously.' Biology, though, was not a 'hands on' subject for him at the time. It wasn't taught at all in school after he was twelve, but Jim made up for this by devouring books, in particular those of J. B. S. Haldane.

Relief from school came in the holidays, when Jim was usually sent away, either to stay with one of his aunts or to a farm willing to take a school-age 'paying guest'. This was largely because it was obvious that the London smog was detrimental to his health, and partly because his parents were too busy to look after him. Lovelock remembers some of these holidays as happy times in the countryside, while others were more like prison camps, especially those in East Anglia where the strong religious views of the farmers made attendance at chapel compulsory and forbad taking a walk, or reading anything except the Bible, on a Sunday. This, of course, simply encouraged his rebellious attitude towards authority. But the biggest holiday influence on Lovelock's development came from the visits he paid to the Leakey family, headed by Aunt Kit and Uncle Hugo.

At a practical level, Hugo Leakey did Jim an enormous favour by teaching him to speak without the unadulterated Brixton accent he had picked up playing around his home. This might not seem like much today, but in the class-conscious 1930s it still mattered, and Lovelock is convinced that he would never have made the most important step of his career, being appointed as a junior scientist at the National Institute for Medical Research, if he had spoken like a south Londoner. More broadly, the Leakeys were left-wing intellectuals and freethinkers, from whom Jim picked up a sympathy for Marxism and the Republican side in the Spanish Civil War, which he managed to combine with being a pacifist. They were also vegetarians; and the whole family delighted in sunbathing in the nude. The Leakeys' 'finishing school' was, says Lovelock, 'the best of my educations'. Hugo's brother, Basil, incidentally, was a professional magician and the inspiration for J. B. S. Haldane's children's book *My Friend Mr Leakey*.

Quite apart from his other idiosyncrasies, being a socialist and a pacifist – he still attended Quaker meetings at the Brixton Friends' Meeting House and met many men who had been conscientious objectors in the First World War – made Lovelock stand out from the crowd at Brixton Grammar School in the 1930s, where most of the boys, and staff, came from more bourgeois backgrounds. The less enlightened of the masters (which means the majority of them) used to poke fun at his beliefs; but this only enhanced Jim's status as the eccentric scientist among his peers, who admired the way he stood up to authority in a way they wouldn't dare. Although there was no bullying among the boys, 'the masters,' says Lovelock, 'were a bunch of sadists.' They seemed to get pleasure from caning young boys on their backsides, although as far as the boys were concerned the potential for pain was greatly reduced by the simple expedient of wearing three pairs of underpants. But the masters also delighted in other forms of humiliation. Lovelock vividly recalls that when he was still in the fifth form he took an examination paper in general knowledge which was intended for sixth formers, who could win a prize; fifth formers could take the paper but weren't eligible for the prize. Lovelock came 'either first or second'. Instead of being congratulated on his achievement, he was hauled up in front of the class by a master who 'harangued him for being "merely a repository of facts, without any real intelligence or ability"'. That kind of put-down gave Lovelock 'an enormous sense of lack of self-esteem' which 'held me back for years'. But some good may have come out of it – as a scientist, in order to avoid such humiliation Lovelock has always been extremely careful before putting forward new ideas to 'make sure that what I was going to do was really right'. That's why he was so annoyed, many years later, when critics of Gaia theory dismissed him as a dilettante who hadn't done his homework.

Although all of this gives an impression of precocious maturity in the teenage Lovelock, in some ways he was much more naive than teenagers today. The school uniform which he was expected to wear until he was sixteen included knee-length short trousers, and his round glasses enhanced the appearance of a scruffy schoolboy to an age where most people today dress and look like adults. Sex was nothing more than a series of 'incoherent fantasies', as Lovelock puts it, until he moved to Manchester in 1939, when he was twenty. But even as a schoolboy, Lovelock enjoyed the independent life. From the age of fifteen, he was able to spend part of his summer holidays each year walking or cycling around England and Wales, covering some sixty miles a day and staying in Youth Hostels. The visits to Wales, in particular, soon developed in him a passion for hill climbing. It was on one of these expeditions, a journey from Kent to Devon in 1934, that he first stopped off at the village of Bowerchalke, in Wiltshire, for tea. On that hot July afternoon it seemed the quintessential English village, nestling in the downs off the beaten track and with its own pub and shop. There and then, Lovelock decided that one day he would live in the village.

In 1936, no longer forced to wear shorts, Lovelock entered the sixth form and passed a 'tolerable' two years preparing for the examination for the Higher School Certificate, the equivalent of modern A Levels. Lovelock describes his younger self as being a 'frustrated physicist'. He really wanted to study physics, and was fascinated by quantum theory, but had great difficulty manipulating the mathematical equations involved. Much later, he learned that he has a form of dyslexia, which means that when he is confronted with an equation he has difficulty knowing which is the left hand side and which is the right-hand side. Although he didn't know at the time why he was having problems solving equations, it was clear that he could not become a physicist, which is why he turned to chemistry.

For some of his peers, the Higher School Certificate was the final step on the road to university; for Jim, the next step in fulfilling his dream of becoming a scientist was a job in South Kensington with the firm of Murray, Bull and Spencer, specialists in all aspects of the chemistry of photography. The key member of the triumvirate, as far as Lovelock was concerned, was Humphrey Murray, a former lecturer at Imperial College, who oversaw the work of the young man and took a tough attitude towards his development. Prospects for advancement within the small firm were non-existent, he warned Lovelock, and it was part of his terms of employment that Jim should attend evening classes at Birkbeck College (like Imperial, part of the University of London) and move on as soon as he obtained a degree in chemistry. The firm would, though, pay the fees for the evening classes, on top of a starting salary of £2 10*s* a week, which soon rose to £3 a week, equivalent to about £10,000 a year today. Interestingly, although he was by then almost 19 and earning a decent salary, looking back on those days Lovelock still refers to his younger self as a 'boy'.

He certainly worked a man's hours, though. He had to wake at 6.30 in order to catch the 7.45 train from Orpington to Victoria and go on by Underground to South Kensington to start work at 9. At 5.30 he left work and travelled to Birkbeck, where he had an hour for supper in the canteen before lectures began. Either lectures or practical classes occupied him from 7 until 9, and he got home at about 11 o'clock, being in bed by midnight. He also had to work on Saturdays from 9 until 12.30. But Sundays were his own – usually occupied in cycling expeditions or walking out into the English countryside, finding somewhere for tea, and running home – typically walking and running for a total of about twenty miles.

It sounds like a peaceful idyll, and most of the time it was. But like most young men Lovelock could get a bee in his bonnet about something he believed in. Many of his slightly older peers felt so passionately about the Republican cause in the Spanish Civil War that they actually went off to fight in Spain, and there seems to be some biological imperative which makes young men become deeply involved in causes of all kinds – an imperative, Lovelock points out, which makes young men so useful to terrorist organizations. As a hiker, one of the causes Lovelock developed an almost irrational passion about was the right to roam, and he hated the landowners who built obstructions across public footpaths. To a young chemist the solution was obvious. He made his own explosives, detonators and all, and from time to time he would set off with them in a bag on the back of his bicycle to blow up the obstructions. 'They never put them back,' he grins, 'it was really most effective.' On one occasion, he blew out 'a few panels' in a high fence running alongside a footpath through the Chevening estate; the fence had been built to protect the privacy of the person who used the house as his country residence – the Foreign Secretary. Lovelock was never caught.

The most important thing that Lovelock learned from Murray was the importance of scrupulous accuracy in his work. Unlike a university student carrying out exercises for practice, it was dinned into him that peoples' lives and jobs could depend upon the results of his chemical analyses. Although it took a long time for Lovelock to reach the standard required, he could never take the easy option of fiddling his measurements to get the right answer, as virtually every student does at one time or another. He also had an early introduction to the hazards of careless treatment of chemicals. After carrying out a series of analyses known as spot tests (because they use just a spot of the material being analysed) Lovelock poured the used material into a beaker that had not been cleaned properly since carrying out a different set of tests using the reagent mercuric chloride. When he saw the mixture produce a growth of delicate crystals that glinted in the light, he picked the beaker up and shook it to see the patterns made as the crystals moved. The resulting explosion left him dazed and temporarily deafened but otherwise unharmed; an apologetic Murray explained that the combination of chemicals was notoriously explosive, but that he had not thought to warn Lovelock because the quantities involved were so small. Lovelock learned from the experience and never suffered another serious explosion during his career; but he wonders whether a modern youngster in an equivalent job would ever get a chance to learn such a lesson, instead being cosseted by safety rules until something went seriously wrong and they were unable to cope.

Work, studying and walking left no time for much of a social life, let alone romance, and in 1938–9 what little social interaction Lovelock enjoyed centred on Birkbeck College. One of the main topics of discussion among the students was the looming threat of war with Germany in the months following the Munich Agreement, which allowed the Nazi regime to take over the Czech Sudetenland. In the spring of 1939, German forces occupied the rest of Czechoslovakia, and the British government announced the introduction of conscription. Most fit young men became liable to be called up for military service, but full-time students were exempt. The part-time students at Birkbeck and similar colleges were incensed by this, since they felt they worked harder than full-time students, had stronger motivation, and would be more likely to benefit from an exemption (and thereby be of benefit to the country) than 'full-time' students who actually spent much of their time socializing and playing games. Articulate and experienced in standing up to authority for what he knew was right, Lovelock became the leader of a protest movement, drawing up a petition which went round all of the evening-class colleges in London collecting signatures before being passed on to the Vice-Chancellor of the University of London in August 1939. Summoned to the Vice-Chancellor's office to explain, Lovelock made his case so well that the Vice-Chancellor agreed to raise the matter with the government. This was a major achievement for a first-year part-time student, just about the most lowly form of academic life; but before the matter could be taken any further Germany invaded Poland and the Second World War began. In anticipation of the inevitable bombing, all London colleges were closed, with full-time students being evacuated to other universities. Part-time students, with their day jobs to consider, had no such option, and Lovelock had to make a decision that would affect the rest of his life.

In spite of his foray into student politics, Lovelock had done well in his end of year exams (a 'far better result than I expected,' he claims) and this opened up the possibility of trying to get a full-time place to study chemistry at another university. There would be no problem about leaving Murray, Bull and Spencer, where he was expected to move on as soon as he had an opportunity for advancement; and it would delay a declaration that he expected to cause him difficulties. Although he had drifted out of the habit of regular attendance at Quaker meetings following the move to Orpington, he still held strong pacifist views and had every intention of registering formally as a conscientious objector when the time came. From listening to the men who had followed this path in the First World War, he expected this to be a difficult experience, being identified as an outcast from society and given unpleasant duties to perform. But since he didn't expect to be called up for at least a year, he might as well study in the meantime.

The place he chose to study was Manchester. This had nothing to do with its superb reputation in science, but was simply because in July 1939 while staying at a Youth Hostel in the Lake District he had met and taken a liking to a girl who, he discovered, was reading chemistry in Manchester. It later turned out that the girl wanted nothing to do with him, but the die was cast. Lovelock's first-year exam results at Birkbeck were good enough not only to persuade Manchester to accept him immediately, under the unusual circumstances created by the war, but to extract a student loan of £60 a year from Kent County Council and a grant of £15 a year from a charitable trust. Lovelock calculated that £75 a year, a bit less than half his income from Murray, Bull and Spencer, would be just enough to live on.

The application procedure was much simpler in those days, when very few people went to university at all. After an exchange of letters with the university, Lovelock was told that if he turned up on such-and-such a date he 'might' be accepted. That was enough for him to burn his boats in London and invest almost half a week's wages in the single train fare north. Lovelock arrived in Manchester with a rucksack, an overcoat and an umbrella. He found accommodation in a cheap hostel until he could settle in student 'digs', and set off next morning to enrol at the university. When he insisted that he should register as a second-year student, the baffled secretary in charge of enrolment sent him to see the Professor, Alexander Todd (a future Nobel Prizewinner and President of the Royal Society, but then only thirty-two). Lovelock made a persuasive case for special treatment, armed with a letter of recommendation from Birkbeck and his outstanding examination results but making no mention of the girl he had met in July. He also, without lying outright, avoided mentioning that he had been studying at evening classes, leaving Todd with the impression that Lovelock had been a full-time student in London. The clinching argument seems to have been his flat statement that he could only afford to study for two years, so he had to start in the second year. Seeing something special about the young man, Todd duly authorized this; but within a month Lovelock was back in his office.

This time, Todd was angry. He had allowed Lovelock special treatment because he saw promise in him, but now he was convinced that the student had let him down by cheating. Lovelock was stunned; he had no idea what had triggered the outburst. Todd explained that it was clear Lovelock had been cheating, because the results of his analysis of a chemical solution were perfect – he had got exactly the right answer. Since nobody could possibly be that good, said Todd, he must have simply copied out the answers and fudged the analysis to match. What was worse, the Professor went on, Lovelock was stupid. If he had only had enough sense to make his answers slightly inaccurate, nobody would have noticed. Todd simply refused to accept Lovelock's explanation that as a professional chemist carrying out work on which people's livelihoods, if not their lives, depended he really had been taught to work to such accuracy – until the Professor came in to the lab and watched while Lovelock carried out another analysis. It was then brought home both to him and to Lovelock how far the standards required of university students differed from those of the professional scientist. And although the experience had been unpleasant to begin with, at the end of it Todd knew that his faith in Lovelock was justified – while Lovelock knew that he had the determination and discipline to achieve his goal of becoming a 'real' scientist. The hallmarks of his life as a scientist – determination, independence, scrupulous accuracy and faith in what he knew to be right – were already clear in the twenty-year-old undergraduate.

When he wasn't bored, chemistry (especially practical chemistry) came easily to Lovelock. When it came to the theoretical side he was used to studying on his own, and found much of the course at Manchester, concerned with the structure of complex organic molecules, uninteresting – not *intrinsically* uninteresting or unimportant, he agrees, just not interesting to him then. So he spent many of the hours he should have been in chemistry lectures attending lectures on history, economics and anything else that caught his fancy. In fact, as Lovelock had suspected, being a 'full-time' student was easier than attending evening classes, and left him free, for the first time in his life, to pursue recreational activities even when he wasn't on holiday. During his time at Manchester he joined the Mountaineering Club and developed a passion for difficult climbs, like other young men ignoring the risks of a sport which killed about a dozen people he knew during his student years. On something of a whim, in spite of his agnosticism he also joined the Catholic Society, which he describes as being then an organization with views similar to those of the liberation theology movement of present-day South America, and so extremely out on a limb in the eyes of the Catholic hierarchy that the Bishop of Salford referred to its members as being 'in the proximate occasion of mortal sin'. It also gave Lovelock the opportunity to meet intelligent and attractive Irish girls (perhaps that was what the Bishop had in mind), one of whom, Mary Delahunty, became his first love.

Although Lovelock had no problem understanding mathematical concepts, his dyslexia and lack of a basic grounding made him very slow when answering questions involving calculations in examinations, so on top of his idiosyncratic approach to chemistry he fell seriously behind in the physics part of his course, which involved lots of mathematical calculation. By the end of his first term, he was running the risk of being asked to leave. Lovelock's poor attendance record at lectures and his patchy exam results could have put him back in Professor Todd's bad books, but a combination of circumstances gave him what Todd thought was an acceptable excuse. At that time, the northern diet, at least in big cities like Manchester, was much more restricted than that in the more affluent south of England, and Lovelock was surprised to discover that there were very few shops where he could obtain fresh fruit and vegetables. There was plenty to eat, but hardly a balanced diet. In January 1940 there was a spell of severe winter weather so extreme that Manchester was essentially cut off from the rest of the country, and for more than a week Lovelock subsisted entirely on baked beans. The result was that he developed scurvy – the Vitamin C deficiency disease historically associated with sailors – and had to be treated at the Manchester Royal Infirmary. Todd (who was, as it happens, an expert on vitamins) was deeply concerned, and imagined Lovelock to be even poorer than he really was, struggling against adversity to fulfil his dream of a scientific education. In Todd's eyes this, along with Lovelock's obvious practical abilities, excused the deficiencies in Lovelock's academic record.

Lovelock sees this as a turning point in his undergraduate career, and for the first three months of 1940 he repaid Todd's faith by working hard – he says, for the only time during his period at Manchester – to make up ground in mathematics. During those months, he also took the step of formally registering as a conscientious objector, even though as a full-time student he was exempt from the call-up. The experience was nothing like as unpleasant as he had anticipated. In March, Lovelock appeared before a tribunal of three judges, who ascertained that he had long held his pacifist beliefs and was a 'fellow traveller' of the Quakers, although not officially a Friend. He was granted unconditional exemption from military service, and he was deeply moved by the civilized nature of the proceedings and the trust placed in him. Soon afterwards, he approached the Manchester Quakers and became a Friend. Typically, he had not done so prior to appearing before the tribunal because he had wanted his case to stand on its own merits, and not to receive the benefit of the favourable attitude of the authorities towards Quakers. Looking back, though, he doubts that he would have taken such a step at all if it hadn't been for the war.

After the intense burst of work in the early months of 1940 and with the dreaded tribunal behind him, Lovelock relaxed and enjoyed life. The romance with Mary Delahunty flourished, and in the long hot summer of 1940, while the Battle of Britain raged over southeast England, Lovelock worked on a Quaker farm some fifty miles north of Manchester, writing to Mary every day. He made such an impression that on his twenty-first birthday the farmer gave him £1, equivalent to about £100 today, enough to pay for a weekend in Blackpool with Mary, where they saw a Spitfire fly past about a hundred feet above the beach – one of the few direct signs of the war, which had as yet scarcely affected Manchester. It was only when he travelled south to Orpington, hitch-hiking a ride in a lorry to visit his parents before the new term started, that the reality of the war was brought home to him. Travelling south through Watford as night fell on 14 September, he saw the sky ahead glowing red from the fires in London, and as they skirted the city to the west the sky on their left was lit up by the fires raging in dockland and the East End.

Back in Manchester, Lovelock found life easy. The main task for third-year chemistry students was to carry out a long series of analyses of various chemical substances, using techniques at which he was so adept, thanks to Humphrey Murray's training, that he completed the entire year's work in the first term. Attending only the minimum number of chemistry lectures and doing the minimum amount of work to ensure he passed his degree left Lovelock time to pay more attention to bacteriology classes, which he chose as a minor option for his degree but found much more interesting than chemical theory – 'it was the one subject I enjoyed and did well at in Manchester.' He also had plenty of time for Mary, and with her help he was able to catch up on much more than his love life. She had an uncle who was a drama critic for the *Manchester Guardian* and could sometimes obtain free seats at the opera, a family who welcomed Lovelock as one of their own and introduced him to 'reading literature for pleasure rather than instruction', and a flat in Manchester into which he soon moved. There was talk of marriage, although that would have to wait until after he had graduated; but Mary had never fallen for Jim quite as hard as he had fallen for her (he recalls that she had turned to him on the rebound from a previous affair), and early in 1941 she broke off the relationship. The last straw seems to have been her first and only meeting with Jim's mother Nell, who assaulted her with 'a lecture on the evils of Catholicism, and the statement that her boy was not to have his career blasted by an early marriage and a string of children'.

The affair ended with a classic 'Dear Jim' letter. Lovelock took his mind off his troubles organizing a student walking party in the Lake District at Easter, and worked just hard enough to scrape a second-class degree. Shortly after the results were announced, he saw Mary for the last time. With nothing to keep him in Manchester, by then the wheels that would take him back south and initiate the next stage of his career were already turning. Lovelock still expresses surprise that he passed his degree at all, having found that a year and a half as a full-time student in Manchester had done wonders for him socially but had added very little to his value as a scientist. He strongly believes that the modern education that requires the brightest scientific minds to spend up to seven years studying for a degree in medicine or a PhD is insane, not least because it shackles the most creative minds just at the time when they are at their most creative, which he sees as a great loss to society. Professionally, he says, the only thing he obtained from his time in Manchester was a job recommendation – but what a recommendation!

During his last term as an undergraduate, like his fellow students Lovelock went for interviews with prospective employers, often as much in the expectation of getting a free lunch as through any interest in the work itself. His outstanding memory of this 'milk round' is a visit to the Manchester firm of Thomas Hedley, part of the Proctor & Gamble empire, where after a series of psychological profiling tests he was solemnly assured that he was not cut out for scientific research, but they would like to offer him a job in marketing. It makes a good anecdote (in 1959, Lovelock used it to introduce a talk he gave as a visiting 'Distinguished Scientist' to a group of Proctor & Gamble chemists in Cincinnati), and clearly they were wrong about the science; but Lovelock's 'marketing' skills certainly came to the fore when he offered the Gaia hypothesis to a sceptical scientific world.

The job that Lovelock actually fell into could have been designed to broaden his education in just the right way to prepare him for his revelation about the nature of the Earth System. In spite of Lovelock's poor showing in exams, he had impressed Todd with his skill as a practical chemist, his determination to stand up for what he knew was right, and in discussions about his Quaker beliefs. Todd was also aware of his unusual student's interest in bacteriology. Unknown to Lovelock, the Professor recommended him to Sir Henry Dale, the Director of the National Institute for Medical Research (an institute funded by the Medical Research Council) and President of the Royal Society, who just happened to be Todd's father-in-law. Sir Henry had asked Todd if he knew somebody to work as a kind of graduate technician, and this seemed a good opening for a conscientious objector with an interest in medical matters who was also a skilled experimenter. The result was that in June 1941, to his surprise, Lovelock was called in to the student office and asked if he would like to travel down to London for an interview at NIMR. It would be the beginning of the second phase of his scientific education and a major step towards the development of the concept of Gaia; a career move he describes as 'absolute blinding luck'. But even before Lovelock started out on the path to Gaia, the concept of the biosphere and the idea of linkages between different components of the living Earth had been aired in unsung scientific speculations – originally, as far back as the late eighteenth century.

3

Gaia before Gaia

The idea of the Earth as a single system, analogous to a superorganism, was already old by the time Jim Lovelock came on the scene, but had been ignored for nearly two centuries. It had been put forward in the mid 1780s by James Hutton (1726–1797), a Scot who was the first person fully to appreciate the way in which geological forces change the face of the planet and the necessary timescales involved. He developed the idea that mountains and continents are constantly being worn away by erosion, but that over the course of geological time worn out continents are replaced by new land uplifted from beneath the sea, recycling the sediments laid down by runoff from the land. At a time when it was still widely believed that the Earth had been created by God in 4004 BC, he saw this as a process involving timescales beyond human comprehension – as he put it, 'we find no vestige of a beginning, no prospect of an end.'

But although Hutton is rightly regarded as the father of modern geology, it is not always appreciated that from the very beginning his vision of the Earth System included life. In the paper from which the quote above is taken, which was read to the Royal Society of Edinburgh in 1785 and published in their *Transactions* three years later, he also said that 'a soil is necessary to the growth of plants; and a soil is nothing but the materials collected from the destruction of the solid land.' In that paper, he described the Earth as 'this beautiful machine', and referred to 'the system of the globe . . . the oeconomy of life and vegetation'. Then comes a bombshell:

> But is this world to be considered thus merely as a machine, to last no longer than its parts retain their present position, their proper forms and qualities?

> Or may it not be also considered as an organized body? Such as has a constitution in which the necessary decay of the machine is naturally repaired, in the exertion of those productive powers by which it had been formed.
>
> This is the view in which we are now to examine the globe; to see if there be, in the constitution of this world, a reproductive operation, by which a ruined constitution may again be repaired, and a duration or stability thus procured to the machine, considered as a world sustaining plants and animals.

The crucial difference between a machine and a living organism, as Hutton realized, is that machines wear out while organisms renew themselves. Hutton was particularly well-placed to appreciate this as he had studied medicine as a young man; he later moved further away from the image of the Earth System as a machine, and specifically made the analogy with a living organism, saying in later presentations to the Royal Society of Edinburgh that the proper study of the planet ought to be physiology. He drew a specific analogy between the circulation of the blood in the human body, described by William Harvey in a book published as long ago as 1628, the circulation of nutrients through the Earth System, and the way water is evaporated by sunlight and falls as rain. In a delicious coincidence, when Hutton was ill and could not read one of those papers to the Society himself, it was presented by his best friend, Joseph Black (1728–1799), who just happens to have been the first person to identify the gas carbon dioxide, then known as 'fixed air' and later to be found to be crucial to the workings of the Earth System.

The reason why Hutton's prescient ideas didn't catch on immediately is that they were so far ahead of their time. Before any true understanding could be developed of the way the living and non-living components of the Earth System interact, scientists had to develop an understanding of each of the components – geology and biology – separately. You couldn't understand the evolution of the Earth System until you understood the evolution of biological systems. And yet Hutton came astonishingly close to the key insight into what biological evolution is all about. In his book *Investigation of the Principles of Knowledge*, a huge, rambling tome which touched on many aspects of science, he wrote:

If an organized body is not in the situation and circumstances best adapted to its sustenance and propagation, then, in conceiving an infinite variety among the individuals of that species, we must be assured, that, on the one hand, those which depart most from the best adapted constitution, will be the most liable to perish, while, on the other hand, those organized bodies, which most approach to the best constitution for the present circumstances, will be best adapted to continue, in preserving themselves and multiplying the individuals of their race.

This is a clear, pre-Darwinian description of what became known as natural selection; and it appeared in print in 1794, fifteen years before Charles Darwin was born. But by then Hutton was in his late sixties, in poor health, and had only three more years to live. For all practical purposes, his ideas about evolution and the Earth as a superorganism died with him.

There is, though, a passing reference to distinctly Gaian ideas in the work of the scientist, explorer and polymath Alexander von Humboldt, who lived from 1769 to 1859, overlapping in time both with Hutton and the next scientist to take up the theme. Humboldt spent the fifteen years from 1799 to 1804 exploring, mostly in South America, and was struck by the interconnectedness not just of living things but of life and the physical world. In the *Narrative* of his travels, published in 1814, he wrote of 'the eternal ties which link the phenomena of life, and those of inanimate nature'. Humboldt's masterwork, *Cosmos*, originally published in five volumes between 1845 and 1861, took its German title (*Kosmos*) from the Greek; the book's translation introduced the word 'cosmos' to the English language. But it was very nearly called *Gaea*; Humboldt changed his mind, as he wrote to a friend in 1834 when preparing the epic, because he wanted a title that would encompass 'Heaven and Earth', not the Earth alone. True to his aim, *Cosmos* ranges over everything from volcanoes and sunspots to the role of marine algae, the origins of language and the invention of the telescope. He aimed to describe 'all that we know today of celestial bodies and life upon the earth'.[1] But his pre-Gaian ideas rather got lost in the mix.

The first real hint of a rebirth of Hutton's idea of the Earth as

1. Letter to Varnhagen von Ense, quoted by Aaron Sachs in *The Humboldt Current*.

a superorganism came exactly three quarters of the way through the nineteenth century; but it was only a hint, and it only concerned the living component of the Earth System. It came from Eduard Suess (1831–1914), an Austrian geologist, as an aside in a book about the origin of the Alps. Suess was born in London, but his father was a German, from Saxony, and the family later moved to Prague for a time before settling in Vienna when he was fourteen. Suess became Professor of Geology at the University there in 1857. He was especially interested in the geological relationship of Europe and Africa, and concluded from his studies of the Alps that they had once been at the bottom of an ocean before being uplifted, so that the Mediterranean Sea is just a remnant of a much larger body of water, which he named the Tethys Ocean. His ideas were largely correct (and very much in line with Hutton's more general speculations), although he did not realize that the old ocean had been squeezed out of existence by the movement of Africa towards Europe, part of the process of continental drift, now known as plate tectonics. Suess was also the first person to conclude from the similarities of fossils found in South America, Africa and India that all three regions had formerly formed parts of a single supercontinent, which he dubbed Gondwanaland.

In 1875, Suess published his book *Die Entstehung der Alpen*, in which he described the structure of the Earth in terms of a series of concentric layers, which he called spheres. Near the end of his book, in a chapter devoted to general observations, he wrote:[1]

> One thing seems foreign on this celestial body consisting of spheres, namely organic life. But this latter is limited to a well-determined zone, at the surface of the lithosphere. The plant, which plunges roots deeply into the soil to feed, and at the same time rises into the air to breathe, is a good illustration of the situation of organic life in the region of interaction between the upper sphere and the lithosphere, and on the surface of the continents we can distinguish a self-maintained biosphere [in his words, *eine selbständige Biosphäre*].

It was this single mention which introduced one of the most important scientific concepts, and it would be decades before the idea of the

1. Our translation. The idea of geological 'spheres' was later extended to the atmosphere, giving us such familiar terms as troposphere, stratosphere and ionosphere.

'biosphere' was taken up in its scientific context and elaborated, by the Russian scientist Vladimir Ivanovitch Vernadskii.

Although Vernadskii's ideas were largely forgotten (or never known about) in the second half of the twentieth century (at least, outside the borders of the old Soviet Union), today he is widely credited as the most important of the 'pre-Gaian' thinkers on the biosphere and the relationship between the living and non-living components of the Earth System. This is probably a fair assessment, although there is little competition for this particular crown, and there has been a tendency among some writers to over-compensate for the previous neglect of Vernadskii by now giving him even more credit than he deserves. This is a source of irritation to Lovelock, who knew nothing about Vernadskii when he wrote his first book, *Gaia: A New Look at Life on Earth*, although he gave him due credit in his later books, such as *The Ages of Gaia*. But although Vernadskii popularized the term 'biosphere' and wrote about what is now called the Earth System in a book published in 1926, 'I defy you,' says Lovelock, 'to find, anywhere in Vernadskii's writings, a clear statement of the importance of feedbacks involving life in maintaining conditions suitable for life on Earth.' And that, in Lovelock's eyes, is the key Gaian concept.

Nevertheless, Vernadskii deserves more than a passing mention. He was born in St Petersburg, in 1863, and studied there and in Munich and Paris before settling in Moscow in 1890 as a Professor of Crystallography and Mineralogy. He was one of the first people to appreciate the possibility of extracting useful energy from radioactive substances such as radium. Towards the end of the first decade of the twentieth century, he began to develop an interest in geochemistry; in 1910 he met Eduard Suess on a visit to Vienna and became interested in his concept of the biosphere, which had by then also appeared in an epic work by Suess called *Das Antlitz der Erde* (translated into English as *The Face of the Earth*). A year later, as part of a political protest against the reactionary policies of the Tsarist regime, Vernadskii resigned his post, along with several other professors, and moved to St Petersburg as head of a new mineralogical laboratory.

Like everybody else in St Petersburg, Vernadskii's life was profoundly affected by the Bolshevik revolution of 1917, and he was also suffering from tuberculosis, another factor suggesting the need for a healthier climate. He moved to the Ukraine, where he helped to set up the Ukrainian Academy of Sciences and established the first 'biogeochemical laboratory' in the world. In the ebb and flow of the Russian Civil War, Vernadskii and his family found themselves at different times on either side of the divide between the Whites and the Reds. By the time the dust began to settle, his son George, who had been born in 1887, had been evacuated (eventually to become a professor of Russian history at Yale), but Vernadskii, his wife Natasha and daughter Nina (then in her early twenties) were moved to Moscow and then to the city that had been renamed Petrograd and would soon become Leningrad. Vernadskii was briefly imprisoned on suspicion of White sympathies, but he was released after a few days because of his value as a scientist. By 1922, he was sufficiently well regarded by the authorities to be allowed to travel to Paris, where he worked at the Sorbonne for three years, carrying out research at the institute founded by Marie Curie and writing his book *Biosfera*, the work for which he is now hailed as a pioneer of Earth System Science.

After his return to what was now the Soviet Union, Vernadskii, in his sixties, worked on various aspects of radioactivity and founded, initially in Petrograd, a group that became the Vernadskii Institute of Geochemistry in Moscow. He was involved in building the Soviet Union's first cyclotron, and the construction of the first heavy water plant in the USSR, both key steps towards the construction of nuclear reactors and bombs. In his late seventies when the Germans invaded the Soviet Union in 1941, Vernadskii and his wife were evacuated with other scientists to Borovoe, in Kazakhstan. His wife died there in 1943, by which time it was safe for Vernadskii to return to Moscow, where he died early in 1945, two months short of his eighty-second birthday.

The best way to get an insight into what Vernadskii offered to the world in his book,[1] and to highlight the key differences between his approach and that of Lovelock, is to listen to the man himself. In the Preface to the Russian edition of the book, Vernadskii writes:

1. Actually a revision for publication in book form of two essays, 'The Biosphere in the Cosmos' and 'The Domain of Life'.

Among numerous works on geology, none has adequately treated the biosphere as a whole, and none has viewed it, as it will be viewed here, as a single orderly manifestation of the mechanism of the uppermost region of the planet – the Earth's crust . . .

As traditionally practised, geology loses sight of the idea that the Earth's structure is a harmonious integration of parts that must be studied as an indivisible mechanism.

So far, so Gaian. But then, in what is clearly a conscious nod to Isaac Newton, Vernadskii says:

I will construct no hypotheses and will strive to remain on the solid ground of empirical generalization.

He returns to this theme in the Preface to the French edition:

The aim of this book is to draw the attention of naturalists, geologists, and above all biologists to the importance of a quantitative study of the relationships between life and the chemical phenomena of the planet.

But:

Endeavouring to remain firmly on empirical grounds, without resorting to hypotheses, I have been limited to the scant number of precise observations and experiments at my disposal.

Even when Vernadskii refers to 'the living organism of the biosphere', he says that it 'should now be studied empirically, as a particular body that cannot be entirely reduced to known physico-chemical systems. Whether it can be so reduced in the future is not yet clear.' These passages highlight the key differences between Vernadskii and Lovelock. As we shall see, Lovelock does construct hypotheses – most importantly, that the interactions between the living and non-living components of the Earth System act to maintain conditions that are beneficial to life – and he does believe that the Earth System can be understood in terms of known physico-chemical systems.

Even so, Vernadskii's empirical insights are impressive. He spells out the crucial point that life on Earth is almost entirely dependent on energy from the Sun, and that the biosphere is a region in which solar energy is used to transform the planet. 'It is evident,' he says, 'that if life were to cease the great chemical processes connected with it would disappear', and no more clays, carbonates or mineral oxides would be formed. 'A stable equilibrium, a chemical calm, would be permanently established,' and the Earth would become an inert, chemically passive planet. 'Life is, thus, potently and continuously disturbing the chemical inertia on the surface of our planet.'

Vernadskii comes closest to the modern concept of Gaia when he emphasizes that 'the gases of the biosphere are identical to those created by the gaseous exchange of living organisms.' But nowhere in his work is there any suggestion that the living organisms regulate the composition of the atmosphere for their own benefit. Rather, 'living matter may be regarded as a special kind of independent variable in the energetic budget of our planet.'

Vernadskii's book appeared in Russian in 1926. Astonishingly, this was a year after the publication in the United States of another book which drew attention to the importance of the flow of energy from the Sun through the biosphere in shaping our planet. That book was originally titled *Elements of Physical Biology*, and its author, Alfred Lotka, was the man who Lovelock now regards as his most illustrious predecessor in the quest for Gaia. Vernadskii and Lotka knew nothing of each other's work at the time, and although each achieved eminence for other reasons, through some quirk of history their insights into the workings of the Earth System were largely overlooked for half a century.

Lotka was born in 1880 in what was then Lemberg, in the Austro-Hungarian Empire; it is now Lwiw, in Ukraine. His parents were American expatriates, and he studied at the universities of Birmingham in England and Leipzig in Germany before moving to the United States in 1902 to work as a chemist for the General Chemical Company. One of the most intriguing features of Lotka's life is that he worked for a variety of employers including the US Patent Office, and from 1924 onwards as a statistician with the Metropolitan Life Insurance Company, while carrying out in his spare time the scientific research for which he is remembered (apart from taking time out for an MA in Physics at Cornell University in 1909 and spending the years from 1922 to 1924 at Johns Hopkins University). Another intriguing feature is that the scientific research for which Lotka became well known wasn't his view of the biosphere as a whole, but his work on the dynamics of animal populations – the relationships between predators and prey that can result in cycles of boom and bust in animal populations. These ideas, also pioneered by the Italian Vito Volterra (1860–1940), became important in the modern study of chaos and complexity, and were also incorporated, as we shall see, into Lovelock's work on Gaia; but their broad significance was not appreciated at the time. Lotka retired from the Metropolitan in 1948, and died a year later.

Lotka was a methodical worker, who tells us[1] that he came up with his big idea during his student days in Leipzig, in 1902. Over the course of the next two decades, he published a series of carefully crafted papers developing his theme of the relationship between life, thermodynamics, and the flow of energy through the biosphere, culminating in his book *Elements of Physical Biology*, which was prepared during his time at Johns Hopkins and appeared in print in 1925. The only piece of thermodynamics we have to take on board here is the famous second law, which can be summed up by saying 'things wear out'. Glasses fall from tables and shatter; cars left untended rust away; houses left without maintenance collapse, but piles of bricks left untended never spontaneously re-arrange themselves into a house. Life seems to defy this process – it is thanks to life, and a supply of energy directed by life, that we have glasses, cars and houses in the first place. And a living organism taking in sustenance from outside sources and growing seems to create order out of disorder. The natural tendency in the Universe at large is for disorder to increase (scientists measure disorder in terms of entropy, so the natural tendency of the Universe is for entropy to increase); but life reverses the process, at least temporarily (entropy is decreased in the immediate locality of life). Lotka was not the only person to appreciate that this is only possible because life on Earth feeds off the energy from the Sun. But he also appreciated that what matters is the *flow* of energy, from the Sun out into the cold of space; the Earth intercepts this flow of energy, and thanks to photosynthesis some of the energy is captured into living systems and flows through them as plants are eaten by animals

1. In the Preface to *Elements*.

which are eaten by other animals, and so on. Most important of all, Lotka tried to quantify this process and work out how thermodynamics affects evolution and determines the amount of life on Earth and its nature.

We don't need to plough through all the mathematics, though, because Lotka's key insight is summed up on page 16 of his book, where he explains why it is valid to study the biosphere in terms of thermodynamics. He says:

> The several organisms that make up the earth's population, together with their environment, constitute one system, which receives a daily supply of energy from the Sun.

And just in case anyone hasn't got the message, he stresses it in a footnote:

> This fact deserves emphasis. It is customary to discuss the 'evolution of a species of organisms'. As we proceed we shall see many reasons why we should constantly take in view the evolution, as a whole, of the system [organism plus environment]. It may appear at first sight as if this should prove a more complicated problem than the consideration of the evolution of a part only of the system. But it will become apparent, as we proceed, that the physical laws governing evolution in all probability take on a simpler form when referred to the system as a whole than to any portion thereof.
>
> It is not so much the organism or the species that evolves, but the entire system, species and environment. The two are inseparable.

No wonder Lovelock likes Lotka! Another reason why he approves of the book is that Lotka uses a systems approach, like an engineer, to quantify his assertions; but we'll stick to the basics.

Here's an example of how the whole system is simpler than its components. Lotka sees evolution as the story of a system undergoing irreversible changes. The ancestors of the whale were fish who evolved into land animals, but when those land animals returned to the water they didn't 'unevolve' back into being fish, but continued to evolve into new forms adapted for life at sea. This is important, because this irreversibility, implying a direction of time, is a key feature of thermodynamic systems, and is associated with the increase of entropy. In this sense, although an individual living thing seems to defy the second law, the evolution of the biosphere is very much in tune with it.

In a paper published in 1922, Lotka spelled out that, 'in the struggle for existence, the advantage must go to those organisms whose energy-capturing devices are most efficient.'[1] And that 'natural selection tends to make the energy flux through a system a maximum,' although the extent to which selection can achieve this depends upon the extent to which random variations, the other key component of Darwin's theory, produce new varieties of life on which selection can operate. 'The influence of man,' says Lotka, 'seems to have been to accelerate the circulation of matter through the life cycle.'

This becomes a recurring theme of Lotka's book, where, in among the maths and physics, he highlights the way fuel and mineral reserves have been built up over geological time by processes involving life, and are now being consumed by human activities far faster than they can be replaced. He comments: 'The great industrial era is founded upon, and at the present day inexorably dependent upon, the exploitation of the fossil fuel accumulated in past geological ages.' But he is remarkably sanguine about future prospects, and says (presumably with nuclear energy in mind) that:

> We have every reason to be optimistic; to believe that we shall be found, ultimately, to have taken at the flood this great tide in the affairs of man; and that we shall presently be carried on the crest of the wave into a safer harbor. There we shall view with even mind the exhaustion of the fuel that took us into port, knowing that practically imperishable resources have in the meanwhile been unlocked, abundantly sufficient for all our journeys to the end of time.

Lotka did, though, have something of a blind spot when it came to the greenhouse effect. Although he referred to the work of Arrhenius and others, he saw the sea as 'a vast equalizer', absorbing most of the carbon dioxide put into the air by human activities, and said, 'as to which side of the account shows a net balance, in the carbon cycle, we have no certain knowledge.'

1. Food is energy, so an energy-capturing device might, for example, be the teeth of a carnivore.

Another of the researchers who Lovelock cites among his eminent predecessors also looked at the whole system of ORGANISM + ENVIRONMENT, but on a more local scale. He was Evelyn Hutchinson (1903–1991), who is sometimes referred to as 'the father of ecology', although his own view was that that title should be reserved for Charles Darwin.

Hutchinson was born in England and studied at Cambridge University before working at the University of Witwatersrand in Johannesburg. He moved to Yale in 1928, and stayed there, eventually becoming Sterling Professor of Zoology, until he retired in 1971; he was a contemporary at Yale of George Vernadskii, Vladimir's son, and helped to get an article by the elder Vernadskii published in *American Scientist* in 1944.

Hutchinson's main area of interest was the ecological study of lakes and ponds (part of the branch of science known as limnology), and he and his students carried out a long-term study of the environment of a small body of water in New Haven, known as Linsley Pond. But although this pond is small by global standards, Hutchinson had broad interests in geochemistry, biology and oceanography, so that he always had an eye also on the big picture. He was always interested in the relationship between the biosphere and the chemical composition of the atmosphere, and as early as 1954, in his contribution to Gerard Kuiper's book *The Earth as a Planet*, he wrote, 'in view of the suboxidised state of the Earth's crust, the occurrence of molecular oxygen in the terrestrial atmosphere is remarkable.' It would be another decade, though, before James Lovelock took this simple observation to its logical conclusion.

Linsley Pond turned out to be an ecological world in miniature. It is a feature known as a kettle hole, a freshwater lake formed when a large block of buried ice melted as the glaciers retreated at the end of the latest Ice Age. Oval in shape, the lake is about 800 metres long, half as wide and with a maximum depth of some 15 metres. By studying every aspect of its biology, chemistry and geological history, Hutchinson's team were able, over several decades, to build up an understanding of the principles that control the behaviour of living communities.

The key insight that emerged from this work was exactly in tune with Lotka's theoretical modelling. It can be seen most clearly by looking at the different varieties of phytoplankton, the tiny organisms at the bottom of the food chain, that live in the lake. Why, Hutchinson asked, should there be different varieties of phytoplankton in the lake at all? If you make the naïve assumption that the pond is a sealed world in miniature, essentially uniform and in equilibrium, then evolutionary theory suggests that one species of phytoplankton should be best suited to the conditions and would out breed all the opposition. In fact, many different species of phytoplankton thrive in the pond, although they are all competing for the same nutrients. The resolution of the puzzle, as Hutchinson explained in an article in the *American Naturalist* in 1961, is that the pond is not in equilibrium but is constantly changing, both with the annual cycle of the seasons and from year to year. Sometimes one species is favoured and starts to dominate; but before it can wipe out the competition, conditions change and another species is temporarily better fitted to take advantage of them. Hutchinson moved from thinking about closed systems in equilibrium to thinking about open systems, which are not in equilibrium but are maintained by a flow of energy (in the form of nutrients, or sunlight, or whatever) through the system. These open systems encourage the evolution of diversity,[1] as the changing conditions never allow one species to become perfectly adapted and wipe out the opposition. There is a clear and exact analogy with the way the biosphere as a whole is maintained in a non-equilibrium state, an open system with considerable biodiversity, by the flow of energy through the system from the Sun.

Hutchinson was also an ecologist in the more colloquial sense of the term. In 1967 he represented the US National Academy of Sciences at a meeting to discuss a plan to put a military base on the island of Aldabra, in the Indian Ocean, which has its own unique biota. Although his formal contribution to the meeting was carefully measured and 'scientific', in an informal account of the proceedings he wrote that 'the intended occupation of the island is a sickening and criminal attack on what I would call a natural work of art, and bad as it is in itself, would set precedents that would impoverish the world

1. They may even explain the origin of life, as discussed by us in *Deep Simplicity*.

even more completely and rapidly than is being done. I cannot believe that the people involved wish to go down in history as they well may with the simple epitaph "they saved money".' Happily, the plan was scrapped and Aldabra became a nature reserve. Although Hutchinson cared about small lakes and tiny islands, the overriding, and very Gaian, message from his work is that the whole is greater than the sum of its parts, whether that whole is the ecology of Linsley Pond or the biosphere of Planet Earth.

A similar message came from the work of the American oceanographer Alfred Redfield (1890–1983). Redfield had an impeccable scientific pedigree, his grandfather having been the first President of the American Association for the Advancement of Science. The younger Redfield was born in Philadelphia and studied at Harvard and in Cambridge, in England, and Munich, in Germany. After a brief spell at the University of Toronto, he joined the faculty of Harvard University in 1921, and maintained his association with the university for the rest of his life. But from 1930 onwards his life and work were intimately connected with the Woods Hole Oceanographic Institution, founded that year on Cape Cod. At first, the Woods Hole researchers worked there only in the summer, but from the time America entered the Second World War it became a full-time research laboratory, where Redfield was appointed Associate Director. The work for which Redfield is best known was the discovery that the ratios of the number of atoms of different chemical elements in marine plankton are the same as the proportions in the open sea. This applies particularly for the elements nitrogen, phosphorus and carbon – for every atom of phosphorus there are fifteen atoms of nitrogen and 105 atoms of organic carbon (that is, not counting the inorganic carbon in the form of carbonate in the shells of these creatures). Associated with this balance, about 250 atoms of oxygen are tied up for every single phosphorus atom. To this day, nobody knows exactly how and why this happens, but as Redfield was fond of saying, it shows that 'life in the sea cannot be understood without understanding the sea itself'. And, we might add, vice versa.

The importance of this particular ratio is that the protein which makes up the largest component of single-celled algae needs a lot of nitrogen, while the nucleic acids which carry the famous 'code of life' that makes the cells work needs a relatively large amount of phosphorus. So the nitrogen/phosphorus ratio is a well-defined property of living organisms. Some of Redfield's thinking along these lines was summed up in an article he wrote for *American Scientist*, published in 1958 under the title 'The Biological Control of Chemical Factors in the Environment'. His specific suggestion was that the nitrate/phosphate ratio in the oceans is the same as that in plankton because the plankton determine the chemical composition of the oceans. This was no more than a hypothesis at the time, and the chemistry involved is so complicated that nobody can say for sure even today how that 'Redfield ratio', as it is known, is maintained. But Redfield's hypothesis is certainly an example of Gaian thinking before the Gaia idea itself was formulated.

Redfield suggested 'not only that the nitrate present in sea water and the oxygen of the atmosphere have been produced in large part by organic activity, but also that their quantities are determined by the requirements of the biochemical cycle'. He concluded that 'the quantity of nitrate in the sea and the partial pressure of oxygen in the atmosphere are determined through the requirements of the biochemical cycle.' But, regardless of his conclusions, the most profound comment in that article was Redfield's remark that he approached the study of the biochemical cycle in the sea 'in much the same way as the physiologist examines the general metabolism of an individual organism'.

By the late 1960s, when Lovelock was beginning to formulate his own ideas on what would become, initially, the Gaia hypothesis, several pieces of evidence pointed in the same direction. Lovelock likes to draw attention in particular to the work of the Swede Lars Sillén (who had the rare distinction of having a mineral, Sillenite, named after him). Sillén was interested in the chemical behaviour of systems in solution, and naturally turned his attention to the biggest such system, the ocean. He realized that the chemical composition of the ocean was not what would be expected if the mixture of chemicals in it were left to settle down into equilibrium, just as the mixture of phytoplankton in Linsley Pond is not the mixture you would expect if the pond were in equilibrium. Among other things, he wrote a paper in the journal *Science*, published in June 1967, on 'The Ocean as a

Chemical System'. 'The chemical system that we must consider,' he said, 'comprises not only the ocean itself but also the other large reservoirs on Earth's surface that interact with the ocean: air, sediments and igneous rocks.' Although he didn't mention life explicitly in that sentence, it soon came in to his story as well.

Sillén made estimates of how much material from 'biologic and photochemical reactions', including nitrogen bound up in molecules such as nitrates, got into the atmosphere every million years, as well as calculating the 'budget' for carbonates deposited in the sea. He looked at chemical cycles involving nitrogen, carbon dioxide and eight other compounds and asked how with all this activity going on the ocean could maintain a roughly constant composition, pointing to the role of 'nonequilibrium processes such as photochemical and life processes', and citing the presence of nitrogen in the atmosphere as 'an especially blatant example of nonequilibrium'. Under the conditions existing at the ocean/atmosphere interface, at equilibrium 'practically all [nitrogen] should be present as nitrate ions in sea water rather than as [molecules of nitrogen] in the atmosphere'. This implies that there must be 'a source [of nitrogen] close to the ocean surface; a process, probably biochemical, that transforms nitrate . . . to elementary nitrogen even in the presence of excess oxygen'.

Sillén's overall conclusion was that 'the present composition of ocean water does not result just from blind chance . . . better understanding of the system OCEAN + AIR + SEDIMENTS and of its history will require close cooperation between geologists, biologists, and chemists of various specialties.'

All of this set the scene beautifully for the emergence of the Gaia hypothesis and theory in the 1970s and 1980s – but that has only become appreciated with hindsight. Before we move on to see just how Jim Lovelock's career prepared *him* beautifully to be the right man in the right place at the right time, though, there is one last piece of Gaian 'pre-history' to recount – the story of how the planet itself was discovered to be 'breathing' with the seasons.

This takes us back to the study of the buildup of carbon dioxide in the atmosphere, and back to Eduard Suess's grandson, Hans. Eduard's son Franz (1867–1941) had followed in his father's footsteps to become a professor of geology at the University of Vienna in 1908. Hans Suess was born there the following year, and lived until 1993. He studied physical chemistry at the University of Vienna, and after obtaining his PhD moved to Hamburg. One of his research interests was the chemistry of heavy water, a form of water in which one of the two atoms of hydrogen in each molecule has been replaced by an atom of deuterium, a heavier form (heavier isotope) of hydrogen. Heavy water turned out to be important in the construction of the first nuclear reactors, acting as a moderator which slows down the neutrons released by the reactor and makes it easier for them to react with the uranium nuclei in the reactor. As an expert on heavy water, Suess was involved in the German attempt to develop such a reactor in the Second World War. He was also interested in other aspects of nuclear physics, especially the origin of the chemical elements (we now know they are manufactured inside stars), and this led him to move to the United States after the war, where he worked at the University of Chicago's Institute for Nuclear Studies, later renamed the Fermi Institute. There, he learned about the technique of radiocarbon dating from the pioneer of the subject, Willard Libby. In this technique, traces of radioactive carbon-14 produced by natural processes in the atmosphere and incorporated into living organisms can be used to infer the ages of old material, such as wood. Carbon-14 is also among the isotopes created by nuclear bombs, and research in this area was partly funded by the US Air Force, which was interested in monitoring the fallout from Soviet bomb tests.

Suess became interested in using carbon isotopes to trace the cycle of carbon dioxide through the biosphere, and began by analysing the proportions of different isotopes in tree rings, which could be dated by counting back the number of rings from the present day. In 1955, he found that the tree rings showed that the proportion of stable carbon from the burning of fossil fuel had increased over recent years, compared with the more or less steady production of carbon-14 caused by the impact of cosmic rays in the atmosphere. This was direct proof that carbon dioxide was indeed building up in the atmosphere as Arrhenius and Callendar had predicted.

This work drew Suess to the attention of Roger Revelle (1909–1991), an oceanographer who was recruiting new members for the expanding Scripps Institution of Oceanography in California, where he had been since 1931 (apart from war service), rising to become its director. Revelle was an exact contemporary of Suess, having been born in Seattle in 1909 and obtaining his PhD, from the University of California, in 1936, the same year as Suess. Together, Revelle and Suess established the new scientific discipline of carbon-14 geochemistry, analysing the circulation of carbon dioxide in the atmosphere, in the vegetation on land, and in the ocean. In 1957, they published a paper in the journal *Tellus* which highlighted the buildup of carbon dioxide in the atmosphere caused by burning fossil fuel, and drew attention to the possibility that this might cause global warming. The mistake they made, however, was in assuming that human activities would continue to release carbon dioxide at the same rate as in the 1950s, from which they inferred that the total increase in the atmospheric concentration of the gas would be only 40 per cent; they failed to appreciate the rapid growth in both human population and global industrialization that actually happened. Nevertheless, in his conclusion to the joint paper Revelle said 'human beings are now carrying out a large scale geophysical experiment of a kind that could not have happened in the past nor be reproduced in the future.'

Suess largely moved on to other work, but Revelle continued to encourage the carbon dioxide studies at Scripps. During 1956 and 1957, geoscientists from around the world carried out many studies of our planet as part of what was known as the International Geophysical Year (IGY). These turned out to be particularly timely, because in 1956 the Canadian-born physicist Gilbert Plass had published the results of the first detailed calculations, based on a computer simulation, of the way the atmosphere absorbs infrared radiation. This provided the best basis to that date for relating an increase in the concentration of atmospheric carbon dioxide to a rise in global mean temperatures, and he concluded, in a paper published in the *American Journal of Physics*, that:

If at the end of this century, measurements show that the carbon dioxide content of the atmosphere has risen appreciably and at the same time the temperature has continued to rise throughout the world, it will be firmly established that carbon dioxide is an important factor in causing climatic change.

That, of course, is exactly what has happened, and we know about the rise in carbon dioxide concentration of the atmosphere since 1956 partly because of the International Geophysical Year. We also know that carbon dioxide is not the only greenhouse gas. Methane is a significant contributor to global warming, and the amount of methane in the atmosphere is increasing as a result of agricultural activities – it is released from rice paddies and from belching cows, for example. There are other, lesser contributors as well; together with methane, they provide an additional 'forcing' of the atmosphere towards a warmer state that is about two thirds as strong as the carbon dioxide forcing today. Putting it another way, about 40 per cent of the anthropogenic warming of the world in the early twenty-first century is due to gases other than carbon dioxide.

One of the contributions to the IGY made by Scripps was to begin taking a series of measurements of the concentration of carbon dioxide in the atmosphere on a daily basis.[1] Although this was initially intended as a short-term project, the measurements have continued almost uninterrupted to the present day, because of the enthusiasm and skill of the scientist chosen by Revelle to make the observations, Charles David (Dave) Keeling (1928–2005). Keeling was a young researcher at the California Institute of Technology who had already developed great skill at building his own instruments and measuring the concentration of carbon dioxide in the air at various places – a research project he chose, as much as anything, because he loved the outdoor life. The place chosen to make the observations was the top of Mauna Loa, in Hawaii, just about as far as it is possible to get from any sources of industrial pollution. Keeling determined to make the observations with the greatest precision and the greatest accuracy possible, and lobbied hard for better (which meant more expensive) equipment to do the job – which turned into a lifetime career for him. The data obtained in connection with the IGY were so precise, thanks

1. Observations of the carbon dioxide concentration had been made before, of course, but only on an *ad hoc* basis; the importance of the new study was that it introduced *continuous* monitoring of carbon dioxide in the air.

to Keeling's search for perfection, that they were able to show an increase in carbon dioxide concentration between 1958 and 1959, the first two full years of the Mauna Loa observations.

In the longer term, these measurements show two things. Taking the average over each year, there has been a steady rise in the concentration of carbon dioxide in the atmosphere, from a pre-industrial level of 285 parts per million (ppm) by volume to 315 ppm in 1958 and to more than 385 ppm today. But superimposed on this steady upward rise is an annual cycle, or rhythm. Because there is more land in the Northern Hemisphere of our planet than in the south, there are more land plants in the Northern Hemisphere than in the south. Each northern spring and summer, the growth of plants takes carbon dioxide out of the atmosphere, and each autumn and winter the decay of plant products such as leaves releases carbon dioxide back into the atmosphere. The result is that the Mauna Loa record of atmospheric carbon dioxide (now known as the Keeling Curve) shows a rise and fall which reflects the breathing of the biosphere. Earth is, indeed, the breathing planet. The Keeling Curve is now as iconic a symbol of the living Earth as the classic picture of our blue planet obtained by the Apollo astronauts.

As James Hutton, a man trained in medical matters, had realized, the proper study of a breathing planet is physiology. The stage was now set for another man trained in medical matters, James Lovelock, to come to the same realization in a modern context. But just how Lovelock became a medical man was a result of another twist in his personal tale.

4

A Medical Man

By the time Lovelock was finishing his degree in Manchester, in the early summer of 1941, he had turned down several job offers (including Proctor & Gamble's bid to make him a marketing man), and been turned down by the chemical giant ICI. He was 'getting near the end of the list' when Professor Todd received a letter from his father-in-law saying that the National Institute for Medical Research needed someone who could help out in a general way – 'it wasn't any very deep stuff' – and did Todd know a student who might do. A conscientious objector with an interest in bacteriology and expertise as a technician in the laboratory seemed to fit the bill nicely.

Lovelock remembers vividly his interview in Hampstead with Robert Bourdillon of the NIMR and 'a sidekick', both Oxford men of the old school who 'thought nothing existed outside Oxbridge and everyone else were peasants'. They didn't expect much of Lovelock except for an ability to act as a glorified lab assistant, but were clearly concerned about his conscientious objection – Bourdillon had flown with the RFC in the First World War and been decorated for bravery. But when the interview turned to hobbies, and Lovelock mentioned his love of mountaineering, the atmosphere changed. 'His face absolutely lit up, because that was one thing where there was no scope for cowardice.' Bourdillon had no time for cowards, but he could respect a man of principle who stuck up for what he thought was right. 'I was in.' It was the beginning of what Lovelock now looks back on as a 'twenty-year apprenticeship' for his life's work.

The work he was recruited to assist with concerned the way infections spread and how they could be controlled. Apart from the direct threat of falling bombs, one of the greatest fears of the authorities was the possibility of an outbreak of an epidemic in the city, where people crowded together in bomb shelters and underground railway stations to escape the Blitz – the powers that be knew only too well that more people had been killed in the 'flu outbreak of 1918 than had died in the fighting in the First World War. Conditions in some of the deep shelters, where Lovelock collected samples, were so bad that a cigarette would not burn, and Lovelock recalls being struck by the discovery that 'combustion is more sensitive to oxygen starvation than we are', an early stimulus for his interest in the relationship between human beings and the environment. In one of the shelters, the oxygen concentration was down to 13 per cent, and 'people were screaming their heads off because they couldn't smoke – it wasn't forbidden, they just couldn't light their matches.' Some of the work also involved studying the common cold, and how to restrict its spread; this might seem an odd subject to devote limited research facilities to in time of war, but Lovelock points out that an outbreak of colds among bomber crews flying unpressurized, unheated aircraft at altitudes of 20,000 feet could make it impossible for them to operate effectively, if at all, with blocked sinuses and Eustachian tubes and with soreness around the mouth making it impossible to wear oxygen masks.

One of the well-documented consequences of living with the constant threat of death and with many social conventions broken down was an increase in casual sexual activity, itself another cause for concern for the authorities, worried about the spread of venereal diseases; the 22-year-old Jim Lovelock was far from sexually inactive. Working in the lab by day, taking air samples from shelters at night, and 'making love with a friendly nurse from Guy's Hospital in the first-aid room at every opportunity', he 'rarely found any time to sleep'. His colleagues at the Institute became very concerned about his health, and decided that he ought to settle down with a nice girl and get married. The girl they chose was Helen Hyslop, the receptionist at NIMR, who had, as it happened, been the first person Lovelock had met when he walked through the door. Encouraged by his friends, and attracted by the thought of a real home life, Lovelock began a 'low-key relationship' (by which he means, no sex) with Helen, which led to their marriage on 23 December 1942. 'It was,' he says, 'not exactly a mistake, but not something that would have happened in normal circumstances – a kind of arranged marriage.' The couple were fond of each other, if not madly in love, and began a very British type of marriage in which there was never any thought of divorce, and everyone got on well enough, but they were really just good friends. After a brief honeymoon in the Lake District, the couple found a flat on the edge of Hampstead Heath, a short walk from the Institute, and the trysts in the first-aid room came to an end.

About the most exciting thing that happened to Lovelock during this period was his duty as a firewatcher at the Institute one night each week. The firewatchers had to stand by ready to extinguish any incendiary bombs that fell in their vicinity, so were exposed high above the ground during bombing raids. This could, naturally, be both exciting and scary; but it had one intriguing side-effect. Lovelock found that after the danger had passed, 'in their post-fear relief normally taciturn senior scientists seemed to feel a kind of confessional need to talk, and would pour out the secrets of their trade to me.' Lovelock drank it all in. 'What I learned on that balcony above London prepared me for a life as a cross-disciplinary scientist.'

By contrast, the placidity of Lovelock's marriage and the quietness of his home life provided a stable background and enabled him to devote himself to his work, which naturally also forms the focus of our attention here. It's striking that what became the basis of the first scientific paper based on his own original ideas was a typical piece of iconoclasm, jumping off from Lovelock's reluctance to accept the received wisdom of his day. Looking back, he says that, 'the whole key to my work, throughout my career, was that whenever they started saying something was the standard wisdom, I started saying "it can't be".' In this case, the received wisdom was that when disinfectants were sprayed in the air, the droplets of spray bumped into bacteria and killed them. But nobody had calculated the probability of this happening, from the known sizes of the droplets and the bacteria. When Lovelock did the calculation, he found that the chance of a collision between a droplet of disinfectant and a bacterium was so small that the bacteria would survive for at least a day. He decided that what must really be happening was that the disinfectant evaporated from the droplets in the air, and then condensed directly onto the bacteria. Nobody believed him at first – 'after all, I was only a kid of 22.' But it was easy to test this idea by experiments with two disinfectants, one of which was non-volatile and largely ineffective, while the other, lactic acid, was highly volatile and 'killed bacteria like the clappers'. He was right.

Lovelock's discovery was published in a scientific paper in *Nature* in 1944,[1] at the insistence of Bourdillon, in order to make the idea public and prevent anybody patenting the idea and making money out of it. This altruistic but misguided attitude affected several of Lovelock's discoveries in the post-war years, and many other discoveries or inventions which could have made money for Britain at a time of economic hardship; it was, says Lovelock, 'a relic of the nineteenth-century attitude which looked down on "trade" as something grubby and amoral'. The Americans in particular had no such compunctions and happily made money out of things the British then saw as beneath their dignity.

In 1945, Lovelock came up with his first invention with an immediate commercial potential. His colleagues were always having trouble trying to write on Petri dishes and other damp glassware with the standard wax pencils supplied, so he came up with a very simple procedure for adding a little detergent to a coloured molten wax mixture before casting it into sticks to make crayons that would write on cold, wet glass. 'They were a roaring success,' he says, 'and I couldn't keep up with the demand from our staff and all their friends in hospitals around London.' Once again, he was told to publish the formula in *Nature*. Within a month, he received a letter from an American company asking to buy the patent; he had to confess that there was no patent, and that the Americans, or anyone else, were free to use his invention commercially.

By then, Lovelock had relinquished his status as a conscientious objector. In 1944, he decided that enough was enough; 'once you started thinking about the merchant sailors bringing the food in, who faced abominable hazards, it seemed so unreasonable.' He was quite prepared to be called for military service, but somewhat surprised to be allocated for training as a medical orderly – logical enough from

1. His name had actually been on a couple of earlier papers, but this was the first for which he was the instigator of the work.

the Army's point of view, but surely not the best use of his talents. When he received his orders, Lovelock tendered his resignation from NIMR; but Sir Charles Harington, who had taken over from Sir Henry Dale as Director, immediately rejected the whole idea. He told Lovelock that the work he was doing at the Institute was far more important, and arranged an exemption on that basis. Ironically, the erstwhile conscientious objector was saved from active service because he was too valuable to the war effort.

There is no doubt that Sir Charles was right. Apart from his work on infectious diseases, Lovelock's most valuable contribution to the war effort came from investigations of the effects of heat radiation aimed at developing techniques for protecting soldiers from burns. One series of experiments involved measuring the burning effect of the heat from flamethrowers, and finding suitable protective materials to shield people from the heat. Together with his colleague Owen Lidwell, Lovelock was instructed to assess the burning effect of heat radiation by exposing shaved (but anaesthetized) live rabbits. 'This was more than we could stomach, so we both came to the conclusion that we would have to burn ourselves.' At first, the experience was 'exquisitely painful'. But after a week or so of regular exposure, to their surprise 'both of us found that the pain decreased, and became more a sense of pressure, even though our skin was burning'. Lovelock has no firm explanation for the phenomenon, but speculates that the excitement and concentration on the scientific work may have caused a release of natural painkillers, endorphins, in their bodies. When discussing his response to 'cruel animal experiments' with us, Lovelock distanced himself from the 'bio activists' who 'terrorize the medical scientists'. His wartime heat radiation experiments were necessary to save human lives, but they didn't have to be carried out on rabbits if human volunteers were available; by the same token, he suggests that 'to offer themselves as volunteers might be quite a useful gesture for the activists to make.'

Not that Lovelock was sentimental about the laboratory animals. At a time of severe food shortages, 'the lab was a great source of food. We often ate Belgian hares that were experimental animals sacrificed for specific organs such as their pituitaries.' Fair enough, but Lovelock's rational approach to the shortage of protein went even further.

The lab work also involved using human blood, kept in a refrigerator, which he was supposed to discard after thirty days by pouring it down the sink. 'This seemed a terrible waste of protein, and some time in 1942 or 1943 I took some home to try mixed with dried egg for omelettes. They were quite palatable – as far as anything made with dried egg was palatable. I suppose that makes me a cannibal.' To put this in perspective, at the time the meat ration was 12 ounces per person *per week*, and if fresh eggs were available at all, each person was allowed one a week.

One slightly surprising outcome of the heat radiation work was the discovery that the best protection against the radiant heat from a flamethrower was an ordinary army blanket soaked in water. 'If you were covered in a wet blanket, you could be fired at by a flamethrower and you wouldn't be hurt.' Even after the water evaporates, the hairy fibres of the wool char on the outside, making a protective coating of carbon which resists the progress of the heat. 'Even a dry blanket was more effective than asbestos. We shouldn't really have been so surprised,' Lovelock says now, 'since animal hair must have evolved to give some protection against the heat from forest fires.' This work was important in the run up to D-Day on 6 June 1944, where as well as the heat from flamethrowers, crews of rocket-launching landing craft were exposed to the backblast from the rockets as they were launched. Three months after D-Day, on 16 September 1944, the Lovelocks' first daughter, Christine, was born.

This work also got Lovelock involved with one of the most bizarre ideas of the Second World War. An organization called the Petroleum Warfare Board, involving all the big oil companies, came up with the idea of setting the seas on fire, using petrol from underwater pipes, to foil any threatened invasion. It was, says Lovelock, 'an example of oil companies going mad', but Winston Churchill was very taken with the idea and test systems were actually built, including one at Studland. 'We towed a wooden boat through it, and demonstrated that it would hardly burn the boat, let alone any soldiers inside. It was utterly useless.' The only argument in favour of the scheme was the psychological effect of the flames on the invaders – but what of the psychological effect on the defenders if they saw the invasion fleet sailing unscathed through the flames?

Lovelock was later asked by Bourdillon to devise something that would detect heat radiation at three levels – corresponding to first, second and third degree burns. He was given 24 hours to do the job, and struck lucky when he discovered by trial and error that a sheet of paper painted with gas-detector paint would do the job. This paint was available everywhere at the time, and was designed to change colour from green to red if exposed to mustard gas. 'To my amazement, it was just right. For first degree burns it went blush pink, for second degree bright red, and for third degree yellow.' Bourdillon was able to take this information to a key meeting, about which Lovelock knew nothing at the time. It was only much later that he learned that the 'detector' was needed in connection with the atom bomb tests.

When the atomic bombs dropped on Hiroshima and Nagasaki brought the war to an end, Lovelock went back to his work on air hygiene, and was moved to the London School of Hygiene and Tropical Medicine, where he was encouraged to spend a year putting this work in order and writing it up as a PhD thesis; during this year, on 26 February 1946, a second daughter, Jane, was born. With a fresh PhD in Medicine from the University of London and a growing family, Lovelock was offered an opportunity to transfer to the Harvard Hospital in Salisbury on a salary of £600 a year, in the virus division of the Medical Research Council (MRC), working in particular on the common cold. He leaped at the opportunity, not because of the salary or because it was a tenured position (if he wanted, a job for life), but because it provided a chance to get out of war-torn London and live in the country, a much better environment in which to bring up two daughters.

The move to the Common Cold Research Unit took place in September 1946, to a centrally heated staff flat at the Harvard Hospital site on the edge of the Wiltshire countryside, with an entire lab allocated for Lovelock's use and medically trained colleagues close by – Lovelock worked particularly closely with Edward Lowbury and Keith Dumbell. The hospital, built with funds provided by Harvard University as a contribution to the war effort, was well-equipped, and provided the base for an 'idyllic' five years. This included an early chance to revisit the village of Bowerchalke, just twelve miles away, once again on bicycle but this time with Helen, and to revive Lovelock's dream of one day settling there.

In 1949, the cold research took Lovelock, some colleagues and some student volunteers to an uninhabited island off the north coast of Scotland to see if isolation from the rest of the world affected their resistance to the cold virus. Locals on the nearby mainland were naturally intrigued, and refused to believe that so much effort was being expended on research into colds. To keep them quiet, Lovelock pretended to confide, in strictest secrecy, that this was just a cover story, and they were really looking for uranium. Less than five years after Hiroshima, this was seen as an entirely plausible explanation, and there were no further questions.

Back at Harvard Hospital, the work also gave Lovelock a chance to hone his skills as an instrument maker. Among other devices, he designed and built anemometers to provide sensitive measurements of wind speed, in order to test the widely held belief that draughts cause colds. One of these instruments used an ionization technique, and could measure air velocities as low as five millimetres per second; more importantly, as we shall see, it was the first of a line of sensitive instruments using ionization devices, in which atoms are stripped of electrons to leave them with an overall positive electric charge, to enable subtle measurements of traces of chemicals in the air.

Another part of the common cold myth was that the fine airborne particles of a sneeze were the main 'pathway' for the spread of infection. Lovelock monitored the way such droplets were spread by using a dropper to put fluoresceine dye in the nostrils of volunteers with colds, and then illuminated their surroundings with ultraviolet light which made the dye fluoresce. Sure enough, it turned out that only tiny traces of the secretions were spread in this way; much more was spread by direct contact from shaking hands, or by a cough into someone's face. This led to a classically simple experiment to check the findings. Two groups of volunteers, one with streaming colds and the other uninfected, were placed in a room, separated from each other only by a blanket stretching from one side of the room to the other. Air could circulate freely over the blanket, and was kept stirred up by fans so that any fine particles sneezed or coughed out by the infected group would spread to the uninfected group. None of the uninfected volunteers caught colds. In the second part of the experiment, the blanket was taken away, and a new group of volunteers was encouraged to mingle with the same group of cold 'donors', eating lunch with them and playing cards and other games. The results were eventually published in the *Lancet* in October 1952.[1] The 'nasal secretions', as Lovelock puts it, were spread from the noses of the cold sufferers to their hands and the cards, then from the cards to the hands and then the faces of the uninfected volunteers. This proved a very effective way to pass colds on – but Lovelock ruefully confesses that nothing much of any real medical value came out of this work.

By then, Lovelock's skill at inventing and operating the ion anemometer to trace air currents in confined environments had already led to his first big adventure – a voyage to the Arctic in the aircraft carrier HMS *Vengeance*. Almost as soon as the Second World War ended, the Soviet Union had gone from being our glorious ally to our potential enemy, and the Royal Navy wanted to test the possibility of operating aircraft carriers in the Arctic in winter; *Vengeance* was picked for a trial voyage to take place early in 1949. One of the objects of the exercise was to test how the health of the crew would be affected by operating under battened-down conditions with re-circulated air at different levels of humidity, so the Navy was looking for a scientist to make air hygiene measurements during the voyage. Lovelock knew nothing of the plan, until he attended a committee meeting at the MRC headquarters in London to discuss a 'dull and pointless' project to monitor air hygiene in schools. One of the committee members was also on the Royal Naval Personnel Research Committee, and asked if anyone knew of a scientist who might volunteer for the Arctic experiment. Lovelock, who had always longed to go on a sea voyage, leaped at the chance. The MRC were very reluctant to let him go, but the Navy had more influence and easily got their man seconded for the job.

It was an exciting experience for a 29-year-old who had never left Britain before, and it is still fresh in Lovelock's memory – he enthuses over the adventure at length in his memoir *Homage to Gaia*. The ship

1. Lovelock is very proud of the fact that he has published papers in journals 'literally covering the entire range from Astronomy to Zoology'.

sailed from Weymouth in February 1949, and headed north towards Spitzbergen accompanied by the usual retinue of ships escorting a carrier, with Lovelock and an assistant berthed in the officers' quarters and their equipment installed in the sick bay. The work involved endless tedious measurements of the kind Lovelock hated, but it was easy and important, and the tedium of the work was far outweighed by the drama and excitement of the voyage as the ships crossed the Arctic Circle and headed west towards Greenland. As it happened, the winter of 1948–9 was unusually mild in the Arctic, but even so it only proved possible to fly aircraft off and land them back on the ship on a few occasions with relatively calm weather, and even then there was a great cost in both lives and planes. 'I've never seen so many air crashes at close hand as on that trip,' Lovelock recalls. 'It was ridiculous, really.' But he was young enough to be infected by the enthusiasm and bravery of the Navy flyers, and volunteered for another ridiculous experiment. The Navy wanted to find out if it was safe to warm up the engines of their aircraft (still piston engines driving propellers in those days) in the hangar below decks before raising them on to the icy flying deck. One of the concerns was the buildup of potentially lethal amounts of carbon monoxide in the hangar; Lovelock volunteered to make the necessary measurements *in situ*, crouching on the hangar floor with his instruments while engines were warmed up all around him. He found that the concentration of carbon monoxide was actually barely measurable, thanks to the fresh air drawn in to the hangar by the whirling propellers; but 'several members of the flying crew, some of the bravest people I've met, came up to me afterwards and said that because of the risk of fire the experiment was foolhardy and dangerous. I'm glad they didn't tell me before the experiment.'

With hindsight, 'foolhardy and dangerous' seems an apt description of the whole Arctic voyage, but that's the way things were done in the Cold War. Part of the experimental nature of the voyage was to find out how the *Vengeance* fared in loose pack ice. The answer was that a relatively mild encounter with a large ice floe cracked one side of the flight deck just forward of the bridge. In severe storms a few days later, well to the north of Iceland, the extreme motion of the ship began to open up the crack, which started spreading across the deck. Strips of metal were welded across the deck by crewmen roped together and to the ship like mountaineers. Fortunately, the storm abated before the ship broke in half, but clearly the experiment was over. Some six weeks after the voyage began Lovelock disembarked at Rossyth, where the *Vengeance* docked for repairs, laden with Navy issue alcohol and cigarettes. 'News of our ordeal must have gone ahead, for when the customs officer asked which ship we came from, and we replied the *Vengeance*, he just smiled and waved us on our way.'

At the end of 1949, the hazards of working with infectious materials were dramatically brought home to Lovelock. One of his colleagues, who he still declines to name, seems to have had an obsession with dangerous pathogens and a careless attitude to handling them. On the afternoon of Christmas Eve, Lovelock became so ill and feverish that his wife called in the duty physician, who said there was nothing to worry about and left for a holiday in Scotland. By Christmas morning, Lovelock's fever was much worse, and he was suffering 'an intolerable itch coming from a small lesion on my right hand; it was like a black ulcer, about five millimetres across'. When Helen called the local GP's surgery, by great good fortune his holiday stand-in turned out to be the local medical specialist. Just about the first thing he asked when he saw the lesion was whether Lovelock had been working with 'nasty pathogens'. 'Of course not,' Lovelock replied, 'we can't have anything of the kind here. The risk to our common cold volunteers doesn't bear thinking about.' The specialist made no comment, but said he would return as soon as possible with penicillin. By the time he did so, 'the lesion was markedly inflamed and a network of red lines extended right up my arm. I can't remember how many times I was given penicillin – massive doses, two million units at a time – but in a few days the fever abated and the lesion began to heal.'

When the first of his colleagues returned from Christmas leave, Lovelock enquired whether anyone had been experimenting with pathogens. He was astonished to learn that 'anthrax and other pathogens' had been used in the lab to teach technicians bacteriology. One of the technicians even mentioned that on one recent occasion a culture plate had been broken by accident on a lab bench where Lovelock often sat to drink coffee. Then, the man who had been training the technicians came back from leave. 'Shortly afterwards, the first colleague retracted his statement and said that he had been mistaken, there was no anthrax in the lab.' Lovelock put the incident to the back of his mind until, many years later, a radiologist working at a Birmingham hospital died of smallpox. The investigation into the tragedy found that one of the virologists at the hospital had been carrying out unauthorized work on the virus, and had been careless in handling it; the pathogen had travelled from his lab through the ventilation system to infect the radiologist. The man concerned was the same man who Lovelock had suspected of carelessness with anthrax back in the Harvard Hospital days. 'The thought that I should have checked to see if my infection in 1949 was anthrax continues to haunt me. Had it been shown to be anthrax then the carelessness at Harvard Hospital would have been revealed, and the much more serious escape of smallpox virus might not have happened.'

During his time at the Harvard Hospital, Lovelock was also involved in one piece of research which had less dramatic immediate consequences, but of which he is now deeply ashamed and embarrassed. In the years after the Second World War there was a great drive to improve the efficiency of British agriculture and get more food from each acre of farmland. This was responsible for ripping out large numbers of hedgerows to provide bigger fields on which larger herds of cattle could graze more efficiently. Lovelock got involved as an adviser to the Grassland Research Institute, at Stratford-upon-Avon. With his fame as an inventor starting to spread, he was asked to design an instrument that could be carried on the back of a bullock and would monitor the movements of the animals – how much time they spent sitting, how much time they spent grazing, and so on – information which would then be automatically transmitted by radio to a monitoring station. The aim was to find out which breeds would be best suited for the new farming style. Lovelock's skill as an inventor was well up to the task, and at the time he never gave a thought to the wider implications. But he now regards himself as 'one of the vandals who helped to destroy the beauty of the English countryside'.

By the early 1950s, Lovelock felt that he had had enough of virology, and wanted a new challenge. When he made his views known in 1951, the head of the Common Cold Research Unit reluctantly accepted the inevitable, while Sir Charles Harington, still the Director of NIMR, welcomed him back to London as just the man he needed to join an important project already underway at the Mill Hill laboratories in North London. There, a team of biologists and veterinary scientists under Alan Parkes were studying how to freeze and thaw cells and biological tissues without causing them too much damage. This potentially offered great medical benefits, for example in preserving tissue for transplantation, and had commercial implications through the application of such techniques to freezing and preserving material such as bull semen for storage and transportation.

Harington told Lovelock that the biologists on the team were first rate in their own disciplines, but knew nothing about chemistry or physics; he wanted Lovelock to keep an eye on them and make sure they didn't make any mistakes, while being free to carry out his own research, in his own well-equipped lab, on whatever caught his fancy. Harington wanted him to start immediately, and arranged for Lovelock's family to stay at the Harvard Hospital flat until they could find somewhere in London. This meant he would be spending each week with his widowed mother-in-law, Queeny Hyslop, at the bottom of Highgate Hill, and commuting by train back to Salisbury for the weekends. As far as Lovelock was concerned, the arrangement worked well. He seems to have been on just as good terms with his mother-in-law, a woman in the mould of his own Nana March, as with his wife, so there was none of the friction that such an arrangement might have caused. This was, as we shall see, very different from the relationship between Helen Lovelock and her mother-in-law, Nell. Things cannot have been too easy for Helen at this time, either. The Lovelocks' third child, Andrew, was born on 2 November 1951, before they were able to find a house in London. Early the following year, they managed to settle in Finchley, aided in their house-hunting by one of Lovelock's colleagues, Alick Isaacs, the discoverer of interferon. Once in Finchley, Lovelock used to join Isaacs on weekend trips to the Kent countryside in search of flint implements and other relics left by Stone Age man; home and family never loomed large in his life.

The research that caught Lovelock's fancy as he settled in at Mill Hill was his own take on what happens to living cells when they freeze. The received wisdom was that the water in the cells and their surroundings freezes into sharp ice crystals that cut through the cell walls, destroying their structure so that the cells cannot be revived. So, naturally, he started saying to himself 'it can't be.' But he insists that he never set out with an iconoclastic agenda. 'It's just that I always try to solve problems from first principles, and when I do, more often than not I find the conventional wisdom is wrong.'

The way Lovelock starts a scientific enquiry, he told us, is through empathy: 'I imagine myself to be in the situation of my subject, and picture what would be happening to me.' In this case, his imagination led him to conclude that from the perspective of the cell, surrounded by an aqueous medium, the effect of ice forming in the surroundings would be to suck water out of the cell to make ice, leaving behind an increasingly concentrated solution of all the substances usually dissolved in the water in the cell. 'I realized that freezing is the same as drying. So what happens to the cell when the concentration of salt starts rising?'

Lovelock didn't come straight out with this idea. He had already started off on the wrong foot with his biological colleagues, who had got the mistaken impression that he had been sent to spy on them by Harington, and he wasn't going to wind them up with the unproven assertion that all their ideas about ice crystals killing cells by spearing them were wrong. Instead, he carried out a series of experiments, first suspending cells in concentrated salt solutions and showing that the damage done is exactly the same as the damage done by freezing, then freezing cells protected by glycerol and other liquids which prevented the cells drying out and showing that the cells could then be thawed and revived as biologically active entities. The protective power of glycerol had already been discovered empirically, by trial and error studies; but it was Lovelock who showed how the technique worked. It was, he says, 'science by intuition, like most good science. This idea that we know what we are doing and plan everything in advance is foolish nonsense caused by the need to apply for research grants. What all good scientists do is make an intuitive guess about something and spend ages and ages testing the guess.' One of his papers on the subject, published in 1953, became the most cited paper in biology that year. By then, Lovelock's skill (and, surely, his own inoffensive and unassuming friendliness) had long since overcome his frosty reception by the biologists, who now treated him as one of their own.

One side-effect of this research was Lovelock's discovery that almost all living cells are damaged by immersion in water with a salinity greater than 5 per cent. This didn't strike him as a very dramatic discovery at the time, but in the context of Gaia theory it is a striking and curious fact that the salinity of the oceans of the Earth has stayed below 5 per cent for well over three billion years, allowing life to exist and evolve. Why? 'I just don't know,' says Lovelock. 'It's one of the biggest puzzles in the whole Gaia story. Is it just a coincidence, or is this an example of life controlling the environment for its own benefit?'

He may not have realized the importance of ocean salinity in the early 1950s, but Lovelock did already have his own views on what life is all about. He chose to work with red blood cells for his experiments, because they are simple structures with cell membranes just like those of other living cells, but with the advantage that if the membrane is damaged the material that leaks out is bright red. He could measure the amount of damage done to the cells simply by measuring the amount of red pigment leaking out, using a spectrophotometer to monitor the redness of the material. The biologists said that this wasn't a valid approach to the study of living cells, because by their criteria red blood cells are not alive. They are simply bags of haemoglobin, and do not have any of their own DNA, so they cannot reproduce, ergo, they are not alive. But, Lovelock replied, they do metabolize, and they maintain their internal properties by homeostasis. To him, this was – and is – a valid definition of life. Three decades later, he would be having the same argument with biologists about whether the Earth could be regarded as being alive, and would be making the same points, in particular the fact of homeostasis, in support of his position.

The most dramatic work that Lovelock got involved in during his time at Mill Hill also raised questions about the meaning of life. The NIMR team didn't stop at freezing cells, but were able to freeze whole animals – specifically, hamsters – and reanimate them with no apparent ill effects. A hamster could be leading a normal life, trained to run a maze for a reward, then literally frozen solid 'like a block of wood', as Lovelock puts it, then thawed out after half an hour or so to run around happily and remember the route through the maze. Before you get too excited, he also points out that hamsters are just about the largest animals that the technique can work with; the success of the process depends on freezing the whole animal right through to its core, in one go, so that all the life processes stop together. In a larger animal – certainly anything as large as a human being – it takes too long for heat in the core of the animal to escape, so that damage is done to the organs while they are cooling down and short of oxygen. There is no way to cool animals from the inside out, which would make the science fiction dream of suspended animation a reality for human beings.

But there is a way to *warm* animals from the inside out, and making this work was one of Lovelock's contributions to the project. If the frozen hamsters were thawed from the outside, by putting them in warm water, then they didn't recover, because as the outer layers thawed they used all the oxygen in the blood before the heart and lungs started working again. To get round this the researchers found that they could revive the frozen hamsters by putting a hot spoon (very hot – warmed using a Bunsen burner) against the chest of the animal so that the heart thawed quickly. This burned the animals badly, which Lovelock found as repulsive as the idea of burning rabbits rather than his own skin to assess the heat from flamethrowers. So he built what he describes as 'a radio frequency diathermy apparatus' to warm the hamsters' hearts without burning their skin. It was, in effect, a microwave oven, which he built out of parts from an RAF-surplus radio transmitter, purchased out of his own pocket at a cost of ten shillings. Lovelock later used the equipment to heat his lunch, which has led to a myth that he invented the microwave oven. He didn't. The use of microwave radiation for heating was well known in the mid 1950s; 'I *might* have been the first person to cook food in this way,' he says, 'but I'd be surprised if nobody else thought of doing it earlier.'

Lovelock was able to discuss the implications of the hamster experiments for the meaning of life with a Jesuit priest, Father Luyet, who also happened to be a scientist and was the main proponent of the idea that the process of freezing cells caused physical damage as they were speared by ice crystals. When he visited the lab in 1954, the team showed him two frozen hamsters. Both seemed to be dead, and one was sliced up to show that there were ice crystals throughout its organs. The other was revived using Lovelock's diathermy apparatus. What did this have to tell us about the nature of life and death? The Jesuit replied that the question was meaningless, because animals do not have immortal souls. Father Luyet's certainty 'rested entirely upon faith. His words stayed with me, and I recalled them in the 1980s, when biologists told me that the Earth "could not" be compared with a living organism. They were also speaking with the certainty of faith.'

The same year, 1954, Lovelock made his first trip abroad, flying to the United States to attend a meeting in Washington on freezing. Even then, Britain was still in the grip of post-war austerity, and Washington 'seemed like fairyland'. After the meeting, at which his work had made a big impact, Lovelock was 'deluged with job offers', and when he received a Rockefeller Travelling Fellowship in Medicine he decided to spend a year at the Harvard Medical School, in Boston. It wouldn't have been his own first choice, but the MRC insisted that if he was to spend a year away from Mill Hill, Harvard was the place to go – although, of course, his pay would be suspended while he was away. The Rockefeller Foundation provided $3,000 for the year, and Harvard Medical School made a verbal promise to top this up with an additional $2,000, which he calculated would be just enough for the family to live on. But there was nothing extra for travel expenses. In order to get the whole family across the Atlantic by ship, the Lovelocks needed £300, which they raised by selling their house. After paying off the mortgage, there was just enough to get them all to the States, travelling tourist class on the *Queen Mary* – but not enough for a return ticket. With typical insouciance, Lovelock decided to cross that bridge when he came to it, and they set off with Christine, now ten, Jane, eight, and Andrew, aged four. The voyage was like a holiday, with food of a quality and quantity scarcely dreamed of in England at that time; but reality hit hard when they arrived in Boston at the start of the academic year.

Lovelock found for himself the truth of the adage that a verbal contract isn't worth the paper it's written on. The administrators at Harvard told him that because of new regulations introduced by the now notorious Senator Joseph McCarthy, they were not allowed to pay aliens, so he would have to forgo the promised $2,000. Their only response to his pleas was to shrug and say, 'Well, you should have left your wife and family in England,' totally ignoring the fact that they would still have needed something to live on back home. A friend suggested that Lovelock could try selling his blood to the hospital. The going rate was $10 a pint, but it turned out that Lovelock has a rare blood group which made his contributions more valuable. He got $50 a pint, almost as much as a week's allowance from his Fellowship, which with regular donations made up the difference between his Rockefeller income and the money they needed to survive. But there was no leeway to build up a reserve to pay for the voyage home, and by early 1955 he was beginning to get anxious. Lovelock found that the cheapest way to get the entire family back to England would be on the cargo liner *Newfoundland*, which sailed from Boston to Liverpool – but even that would cost £250. Then he saw an advertisement for an essay competition. The CIBA Foundation was offering £250 for the best essay on 'Research in Aging'. The sale of another pint of blood brought in enough to pay for a second-hand typewriter, on which Helen Lovelock pounded out Jim's thoughts on the subject in February 1955. 'I managed to include a reference to Einstein,' Lovelock recalls, 'which I thought was a rather nice touch.' The essay, largely based on his work freezing and drying red blood cells, won the prize, and the Lovelocks sailed in September. By then, Helen was pregnant with their second son, John, who was born on 16 February 1956. Unfortunately, he suffered brain damage caused by anoxia just after being born, but in spite of a difficult childhood survived to manage his life well. In order to cope with his hyperactivity, when John was a child the family had to have a rota system, with somebody always 'on call', night and day, to be in charge of him. He could never have fitted in to an ordinary school, but the education authority in Wiltshire offered to send him to a special school, and Lovelock's friend William Golding suggested a Steiner school at West Hoathly in Sussex, which, says Lovelock, was 'tough, with strong discipline, but good; they achieved great results'. John still lives close to Jim in Devon.

In the year that John was born, Lovelock's father died. His mother, Nell, had never got on with Helen, and because as a child Jim had been largely brought up by Nana March, he had never formed a close bond with Nell himself; the result was that over the next quarter of a century she 'loomed like a dark cloud' over the family.

In exchange for his harsh treatment in Harvard, Lovelock brought to the Medical School the knowledge they needed to develop blood-freezing techniques, without them actually paying a penny for it. He only learned how valuable his contribution had been when it was time to return home, and he was suddenly offered $6,000 to stay for another year. The McCarthy story, he realized, had been a complete fiction. When he angrily turned down their offer, it was increased to $10,000, which had the effect of reinforcing his determination never to darken their doors again, but provided him with an important lesson about the way to negotiate financial terms with American institutions. Lovelock's long-time colleague Lynn Margulis says that he is 'incapable of deception . . . very transparent and open about his own actions', and that he 'hates the academic one-upmanship game'. The Harvard experience reinforced his determination to steer his own course.

By the time he got back to England, Lovelock was again beginning to suffer from academic restlessness. The cryobiology work had reached the point where freezing and reanimating hamsters was almost routine – by now, they could freeze hamsters at −5°C for up to an hour before reviving them – and he had an urge to do more with instrumentation and detectors. 'All the time,' he says, 'I was at heart a physicist, applying physics in a biological context.' The Director had no objection, and Lovelock had a chance to wrap up his work on freezing in a paper presented to the Royal Society early in 1956.

How does this work stand up half a century later? In 2006, Lovelock was elected as a Fellow of the Society for Cryobiology, in direct recognition of his pioneering work in the early 1950s. The President of the Society, Andreas Sputtek, told us that Lovelock's papers on the nature of freezing damage to biological cells are classics which are still read today by every student of cryobiology. Ken Muldrew, a cryobiologist based at the University of Calgary, described the work to us as 'utterly definitive. Lovelock's papers are marvels of clarity and economy, demonstrating clearly that the primary mode of freeze–thaw injury to cells was through dehydration . . . a perfect example of the use of empirical experiment as a "tool for thought".' He points out that cryobiologists spend a long time and a lot of effort 'probing the phase transition that separates the living state from the non-living', and says, 'I believe that Lovelock's early work in cryobiology was pivotal in directing his imagination toward an understanding of what "life" is,' and leading him to the Gaia hypothesis, which Muldrew describes as a 'brilliant unifying framework for biology'. When he first encountered the Gaia hypothesis, Muldrew says, 'I was struck by the similarity between the living planet having to maintain a critical amount of complexity to sustain life, and a cell, when subjected to some harsh physico-chemical insult (such as a freeze–thaw cycle), only being able to repair itself if there remains a sufficient collection of interacting systems.' But that is getting ahead of our story. In 1956, Lovelock still had one more far from uncongenial task to perform for Parkes and his team before he left cryobiology for good.

The obvious dramatic potential of the freezing work had inspired the Director's former secretary, Lorna Frazer, who had left the Institute to become a writer, to come up with a play based on the research, but scaled up to freeze a human being. The play, *The Critical Point*, was about to be produced by the BBC, starring Leo McKern and Joan Greenwood. Lovelock was given six weeks' leave to act as science adviser on the project, ensuring the laboratory scenes were realistic. When he carried out his brief to the letter, everyone was disappointed; a faithful representation of the Mill Hill laboratory wasn't sexy enough for dramatic purposes, and viewers complained that the set wasn't 'scientific'. When a second broadcast of the play was planned, a completely new production in those days before videotape, Lovelock was again consulted, but persuaded to 'let them have their head and dress the set like the control room of a nuclear power station. Everyone was delighted.' The only real contribution he made was to devise an audiotape, using an electronic sound generator, to imitate the dying breath of a murder victim, his death rattle, and his failing heartbeats. To his surprise, the BBC paid £50 for the tape, without being asked; he later learned that it became the inspiration for the establishment of the BBC's Radiophonic Workshop, best known for its production of the *Dr Who* theme.

The very last paper Lovelock produced before leaving Parkes's department concerned the use of a substance known as dimethyl sulphoxide as a protective agent for freezing red cells. He remains particularly proud of this work because he predicted that the properties of dimethyl sulphoxide would make it ideal for the job, and this was then confirmed by experiment. He was not to know at the time that a closely related compound, dimethyl sulphide, or DMS, would later become an important part of the Gaia story; but his next career move, sideways at Mill Hill into the biochemistry department, would be instrumental in setting him on the road from animal physiology to planetary physiology.

5
Inventing the Future

It was in the biochemistry department at Mill Hill, from 1956 onwards, that Lovelock was able to give full rein to his skill as an inventor of sensitive scientific instruments. This changed not just the course of his own career, but the entire future of environmental science, when those instruments proved sensitive enough to detect pollutants in the environment in almost unimaginably small traces; the inventions really made the environmental movement of the second half of the twentieth century possible. But the most important family of these detectors actually harked back to Lovelock's work on the common cold, in 1948, when he had been asked to build a detector to monitor the movement of gentle draughts of air.

Lovelock's sensitive anemometer was based on the idea of using a small amount of radioactive material to make positively charged ions which drift with the air. The radiation from the radioactive source literally knocks negatively charged electrons out of a few atoms, leaving positively charged ions behind; the positive electric charge provides both a 'label' which makes it relatively easy to follow the ions as they are moved around by draughts and, in some applications, a 'handle' by which the ions can be moved by magnetic or electric fields.

One of Lovelock's recurring complaints about modern science is that it is too tightly constrained by 'health and safety' regulations, which make it impossible for his counterparts today to work as effectively as he did in his youth. The ion-drift anemometer is a classic example of what he means. In order to get radium to produce the radiation he needed for the ionization process, Lovelock scraped the luminous paint from the dials of instruments from scrapped war-surplus aircraft – instruments anyone could purchase at ordinary surplus goods stores in London. Nobody bothered about the tiny trace of radium that made the paint luminous. Indeed, like many of his generation, for years Lovelock wore, 24 hours a day, a wristwatch on which the hands and numbers would glow in the dark because they had been painted with a material containing radium; but we are now far too cautious to allow such use of radioactive material. The essentially home-made anemometer, radioactive paint and all, worked beautifully both on land and on the *Vengeance* voyage, except that it was very sensitive to cigarette smoke, and it was also affected by gases known as chlorofluorocarbons, or CFCs, then widely used in refrigerators. 'At the time, this was just a nuisance. I never imagined the way CFCs would affect my life.'

The work that drew Lovelock to the biochemistry department at Mill Hill five years later was chiefly being carried out by the Nobel-Prizewinning biochemist Archer Martin, just nine years older than Lovelock, and his colleague Tony James. The technique they were working on, known as gas chromatography, was so exciting and important that Lovelock had already started collaborating with the team before formally making the move. It is a development from the idea of paper chromatography, memorably demonstrated in school, when a piece of blotting paper or tissue is marked with coloured spots from marker pens and dipped in water. As the water rises through the paper by capillary action, it spreads the different blobs out into the different coloured components the inks are made of. The technique of chromatography, actually using a tube full of powdered chalk rather than a sheet of paper, had originally been devised by the Russian botanist Mikhail Semenovich Tsvet, in the early 1900s, and named by him in 1906; he used it to analyse the coloured pigments from flowers, and chose its name from the importance of colour in this early work. In fact, it can be used, and most often is, to separate out colourless components of a chemical mixture, so the term is rather misleading even in this context, and is even less appropriate when used in the expression 'gas chromatography', but we are stuck with it.

In gas chromatography, a mixture of gases is passed through a tube containing a fine powder (perhaps itself coated with a non-volatile liquid) and as the gases move along the tube different components of the mixture are absorbed by the powder at different places along the tube. If you can measure the amount of each substance absorbed at each site along the tube, you can work out the composition of the original mixture of gases – in particular, if you have sensitive enough instruments you can measure the amount of trace substances in the mixture. The biochemists were interested in analysing things like the gases responsible for giving vapour a scent, or the components of petroleum oils, or the mixture of compounds found in different sorts of biological material. It was Lovelock who turned the technique into a tool for analysing pollutants in the atmosphere.

Lovelock confesses that 'with hindsight, I made too many inventions in those days. I kept coming up with ideas, but I never took time to develop most of them properly.' Even so, several of his detectors became invaluable in gas chromatography and other applications; but we shall only describe the two most important ones.

The technique Lovelock developed used the same basic physics as his ionization anemometer. By 1956, he no longer had to scrape luminous paint off RAF surplus instruments to get a radioactive source, but his approach would still horrify a modern health and safety executive. The radioactive material he worked with, strontium-90, was provided for researchers in the form of a stiff silver foil, which emitted beta radiation (another name for fast-moving electrons). Usually, this would be kept in a lead-lined box or other container to stop the radiation escaping, but Lovelock had to cut it and bend it to fit his detector. 'I worked the foil behind a thick sheet of glass, so I could see what I was doing. The glass stopped most of the beta rays, but not all of them, so I worked out in my head roughly how long it would be safe to work with the material in each session. There were no rules about such things; you were expected to use your own common sense.'

Towards the end of 1956, Lovelock was struggling to construct a sensitive instrument for the needs of Martin and James. He was using a two-stage technique, in which a 'carrier' gas flowing through the instrument was exposed to radiation, and the impact of the beta rays on the atoms of the gas gave some of them extra energy, putting them in what is called an excited state. When one of these energetic, or excited, atoms collided with a molecule of the vapour being analysed, it gave up its energy and knocked an electron out of the molecule. The positive ions produced by this process could then be collected and counted using magnetic fields. The trouble was, Lovelock couldn't find a good carrier gas. The technique had actually been devised by American scientists who recommended using hydrogen or helium. The MRC couldn't afford helium, which was much more expensive in the 1950s than it is today, and even in those days hydrogen, which burns more vigorously than any other substance, was clearly too dangerous to use in an experiment which had to be left running all night. So he was using nitrogen – a nice, safe gas – and essentially getting nowhere.

'I was about to give up,' Lovelock recalls, 'but decided out of sheer bloody-mindedness to try one last run. All the nitrogen in the lab had been used, so I sent my technician to get some more. She came back from the stores with a cylinder of argon instead. The stores had run out of nitrogen, but she had used her initiative and asked me if argon would do. I said to myself, "Why not?"' In theory, the detector should have been slightly less sensitive with argon than it was with nitrogen, but Lovelock felt he had nothing to lose, 'so I connected up the cylinder and started what I thought would be my last run before wrapping that job up. To my surprise, the recorder went off the scale. I had to turn down the sensitivity on everything; the argon amplified the signal thousands of times.' The instrument became known as the argon detector – which still brings a smile to Lovelock's face, since 'argon is the one thing it can't detect.'[1] It was a sensation in the biochemical world, and over the next couple of years he made many trips to other labs, especially in the United States, to describe the discovery. He was offered jobs in America at twice his Mill Hill salary, but wasn't tempted to leave England. Then, the bubble burst when someone else invented an even better instrument, called the flame ionization detector; but the heady experience had sent Lovelock's reputation among his peers soaring.

During the last years of the 1950s, Lovelock also developed his

1. In his key paper describing the instrument, published in 1958, Lovelock couldn't resist a tongue in cheek allusion to the way he came to use argon. 'By a fortunate coincidence,' he wrote, 'argon functions well in the device when used directly from commercial cylinders.'

most famous invention, the electron capture detector, or ECD. The breakthrough, in 1957, he describes as 'the most important event of my scientific life'. Although there were many technical hurdles to overcome before the instrument could be made to work, the underlying science is disarmingly simple. In this case, nitrogen works very well as the background gas, but the active component is not the positively charged ions left behind when beta rays knock electrons out of the nitrogen atoms, but the freed electrons themselves. If the detector is full of nitrogen, all of the electrons released in this way can be attracted to a positively charged electrode, where they produce an electric current which depends on the number of electrons being released. But if a stream of some trace substance such as DDT or one of the CFCs is mixed with the nitrogen gas, the molecules of the 'pollutant' absorb the electrons and the current drops. Almost exactly one electron is captured for each molecule of pollutant present; even better, each kind of chemical compound has a different 'affinity' for electrons, so that the detector works as a conventional gas chromatograph to identify the kinds of molecules present, as well as their quantity. Even in its early stages of development, the ECD could detect the presence of a couple of hundred thousand molecules of DDT in a cubic centimetre of air. This amounts to one tenth of a femtogram of DDT, and one femtogram is one billionth of a millionth of a gram. Since the air we breathe is made up of 78 per cent nitrogen, the technique could immediately be applied to measuring traces of pollutants in the atmosphere. Unlike the argon detector, the ECD has never been superseded, and the best modern instruments can directly detect one part of pollutant in one hundred thousand billion parts of air; because conventional chemical techniques make it possible to concentrate the pollutant at least a hundredfold before the ECD is used, in practice this means that Lovelock can detect a trace of gas as small as one part in ten million billion – 1 in 10^{16}. By 1959, Lovelock had incorporated the ECD in an easily portable analytical device, the most sensitive instrument of its kind; 'It still is,' he adds.

There was one sour note, which still grates with him today. Part of the development work for the argon detector and in particular the ECD was carried out at Yale University, where Lovelock worked for eight months in 1958–9, enjoying another happy sojourn in the United

States with his family. 'Kindness, hospitality and an environment conducive to creative work helped us get over the memories of the difficult year in Boston, but there was one unforeseen consequence.' His colleagues at Yale urged him to apply for a patent on the ECD, which he did, agreeing to share any income with the university. Unfortunately for Lovelock, it turned out that there was an agreement between Yale and government bodies that paid for their research that all patents filed from the Yale School of Medicine became the property of the US Surgeon General's Department. 'I didn't think the ECD would be worth much, so I didn't put up much of a fight. Looking back, I probably could have got at least a share of the royalties if I'd pressed my case. But as it is, I made nothing at all out of the ECD.'

Not that Lovelock felt any pressing need for money, in the late 1950s or since. With the security of a tenured job and a good salary that was guaranteed to increase as he grew older, in 1957 he and Helen decided to fulfil his lifelong dream of buying a cottage in Bowerchalke. They weren't wealthy enough to contemplate the purchase of a second home on their own, but were able to cover the cost of a mortgage for £1,000, half the purchase price of Pixies Cottage, while a colleague at Mill Hill, Thomas Nash, put up the rest. Nash was a wealthy bachelor who treated the Lovelocks as a kind of surrogate family, often going on holiday with them; Lovelock, easy-going as ever, got along well with him, although Helen was never quite comfortable with the relationship. The informal arrangement was that as long as the Lovelocks lived in the cottage they would pay the mortgage, but when they moved on the cottage and any remaining debt would revert to Nash.

The desire to escape to the country may have been stimulated by what Lovelock still describes, more than fifty years later, as 'a really devastating event' – the death of his father from kidney failure in 1956. The death itself was peaceful enough; Tom Lovelock died quietly in his sleep without waking, and both Jim and Nell were at his bedside, at Chelsfield, near Orpington. But both of them were deeply affected by the loss. 'The death of my mother was much less traumatic. She moved to Plymouth to be near us in 1986, when she was 90, and died a couple of years later, in a Quaker retirement home. It happened in the night, and I was informed by a nurse the next day. But it was much less traumatic than the death of my father, partly because she had been ill with dementia for years, and partly because my childhood bond had been to my grandmother, not my mother.'

Although still living in the suburbs of Finchley, which they had no affection for, the Lovelocks now took every opportunity to spend weekends and longer spells at the tiny two-bedroomed refuge. They loved the village way of life, still thriving almost half a century ago, and became increasingly disillusioned with suburbia. Things came to a head not long after the family returned from Yale, where they had lived in the kind of small community they preferred, Orange, just outside New Haven. It wasn't only the suburbs that depressed them; by the beginning of 1960, there was also the baleful presence in Finchley of Lovelock's widowed mother to contend with.

By 1960, the Lovelocks lived in a five-bedroomed house in Finchley, purchased in 1959 for £3,500, only a little more than Lovelock's annual salary of £3,000; he also had income from lecture tours in the US, but as well as paying for the mortgage on the cottage in Wiltshire he contributed 5 per cent of his income to Nell, who lived in a flat nearby, to boost her pension. Unfortunately, Helen and Nell, 'both good, well-intentioned women, but as stubborn as each other', disagreed bitterly about the way the Lovelocks' children were brought up. Jim himself was unhappy with the situation, and increasingly restless at work, where the security and comfort of a job for life began to look more and more like a prison as he approached and passed his fortieth birthday. He had 'a vague urge to do something different – make a complete break with the past. I fantasized about making a living writing science fiction or becoming a freelance consultant on instrumentation. But how could either occupation provide the support my family needed?' There was actually one possibility, although it wasn't tempting enough for Lovelock to make the leap. By this time, the official attitude to patents had changed, and the team at Mill Hill was encouraged to patent their inventions, not for their own benefit but for the benefit of British industry. The researchers were then expected to give free advice to British firms using their inventions, and Lovelock developed a good relationship with one company, Pye, based in Cambridge. He got as far as asking them how much they would pay him as a consultant if he did decide to leave the Institute; the answer was £2,000 a year. 'Enough to get by on, but not enough to make the decision easy.'

What turned out to be a crucial step came in the spring of 1960, following a particularly severe row between Helen and Nell. Christine Lovelock, Jim and Helen's elder daughter, was having exactly the same kind of trouble with her school that Jim had had as a teenager. She got on well academically, but could not see the point of homework, and refused to do it. Her parents saw no reason to force her, but her grandmother, still bitter about her own lost opportunities, 'saw Christine's attitude as an ungrateful rejection of a chance for a better life, and expressed her views forcefully'. Helen Lovelock, a quiet woman who had the British 'stiff upper lip' attitude that made any outward show of emotion distressing, was so upset that the next morning she and Jim decided to move the family to Wiltshire immediately, selling the house in Finchley and looking for a bigger cottage in Bowerchalke. It meant Jim living with his mother during the week, and commuting 95 miles by car each weekend – a daunting prospect in those pre-motorway days, and only made feasible by driving at speeds of up to 90 miles an hour; once free of the suburbs, there were then no speed limits in the countryside. 'I knew perfectly well I wouldn't be able to cope with this way of life for long; I think my subconscious was encouraging me to put myself in a position where I would have to resign from the Institute and do something different.'

When he wasn't working or driving, Lovelock began to drink heavily to ease the stress, and developed a chain-smoking habit. 'By January 1961, I was worn out mentally and physically, and had to spend five weeks recuperating at Bowerchalke. I managed to get fit enough to go back to work, but I knew it couldn't last. I was too tired to make any real plans to get out of the hole I'd dug for myself, and things were complicated by an offer from America.' The US National Institute for Health offered Lovelock a research grant of $50,000 a year for three years, to carry on his work at Mill Hill developing detectors for use in lipid biochemistry (lipids, which contain fatty acids, are among the most important molecules in the human body). On the face of things, this was exactly the kind of commitment Lovelock did not want, but he felt it was his duty to inform the Director. To his relief, the MRC turned the offer down, apparently feeling that there was something 'not quite right' about taking foreign money to fund research at the Institute. A few weeks later, however, in March 1961 another offer arrived from the United States. This time it was one Lovelock simply couldn't refuse, and the MRC couldn't veto.

Lovelock is one of those people who, in a quiet way, seems to get excited and enthusiastic about everything he does. But even in the context of this enthusiastic attitude to life, his excitement at the opportunity to become a space scientist stands out. As a science fiction fan and frustrated physicist, to get an invitation from the fledgling NASA to join the team planning to send spacecraft to explore the Moon was clearly a highlight in his life, not just his career, and his eyes still light up when he describes the moment. 'I was being asked to join in the kind of adventure that just a few years back I had been reading as science fiction. It was literally the stuff of my childhood dreams. I was ecstatic.' Even better, the offer gave Lovelock the perfect excuse to leave Mill Hill. 'For more than a year I'd been trying to find a face-saving way out. Everybody there was so nice, and I knew they wouldn't understand my reasons for leaving. But suddenly I had the answer. They all knew I was mad about astronomy and space. I could quit without causing any offence.'

The project Lovelock was asked to join was the first unmanned American lunar lander series, Surveyor, which paved the way for the manned Apollo missions. The obvious thing would have been for him to work full-time at NASA's Jet Propulsion Laboratory (JPL) in Pasadena; but that would have been too simple. 'I didn't want to swap one institute for another, so I planned to fix up a job as Visiting Professor in Houston, where a friend of mine, Al Zlatkis, worked in the chemistry department. From there, I could apply for a NASA grant to do the detector research JPL wanted.' Things soon turned out even better. A new lipid research centre was being set up at Baylor College, also in Houston, and when news of Lovelock's move reached them, he was offered a post as research professor there on a salary of $20,000 a year, working on just the kind of detector development NASA needed. There was no question of selling his blood on this trip. Over the next couple of years, Lovelock's income from all sources rose to about $40,000 a year, much of which he put aside to enable him to set up as an independent scientist when he returned to England.

With money in the bank, the Lovelock family enjoyed their time in Houston. Once again they crossed the Atlantic by ship (this time, the *Mauretania*), but now with no worries about the return fare, and settled in a new five-bedroomed house with, to Jim's delight, more than twenty species of snake in the garden. 'They included several kinds of rattlesnake, copperheads and others. But they didn't bother us, and we didn't bother them.' They were introduced to the delights of credit cards, virtually unknown in England at the time, and Christine and Jane attended lectures at the University. Christine became engaged to an Arab student from Gaza, but in the end the couple decided that the cultural differences were too great for her to return home with him. The girls' brother Andrew had a less enjoyable time. In those days, the English primary school system was far better than the American one, and Andrew was so far ahead of his age group educationally that he was bored and goofed off school – he couldn't even be put in the same class as his academic peers, because the age difference was so great. He was 'always in trouble', says Lovelock, so after the first year he returned to Bowerchalke, where Helen's sister was living in the cottage, and went back to the village school, eventually gaining a scholarship to Bishop Wordsworth school in the nearby town of Salisbury.

Lovelock himself went native in Houston to such an extent that it almost got him into trouble – on one occasion on a visit to New York he ordered a sandwich at a bar and only realized after he had ordered that the only money he had on him was a $100 dollar bill that he kept for emergencies. When he proffered it and the barman snapped, 'Haven't you got anything smaller?' he replied without thinking, 'We don't use anything smaller in Texas.' Instantly, 'the atmosphere in the bar became frosty, and I was only saved when they realized I was English.'

But although the life in Houston was enjoyable, these were 'the least productive years of my working life'. The Lipid Research Laboratory was superbly equipped, and there were funds to buy any equipment Lovelock wanted; but paradoxically he found this stifling. He was always happiest when building his own instruments out of war surplus equipment and bits other people had discarded. The American way was too easy. Instead of inventing his own instruments to answer specific scientific questions, he spent his time playing with the new toys and finding out what they could do. It was also, he says, 'probably not a good idea, as far as work was concerned, to spend every summer back in England. And then there were my monthly trips to JPL. I'm afraid Baylor didn't really get value for money out of me.'

Although usually Lovelock travelled the 1,700 miles from Houston to Los Angeles by air, sometimes he took the entire family by car, taking two and a half days, or longer if they detoured for a visit to sites such as the Grand Canyon or Meteor Crater. There was no incentive to hurry, because after his initial excitement at being part of a science fiction story, Lovelock found the lunar work at JPL dull. The long-term objective might be to send men to the Moon, but the day-to-day nitty-gritty was an endless round of discussions on the details of gas chromatographs to be used for analysing the composition of the lunar surface. It was only towards the end of his time in Houston that Lovelock's enthusiasm for the space program was revived, when the people at JPL started to turn their attention to Mars, and the search for life on the red planet. Lovelock felt he had provided as much help as he could in terms of lunar chemistry, and had an affinity with the engineers who had the fascinating (for him) challenge of making the instruments to do the jobs the scientists – in particular, the biologists – wanted. With his unusual background, he found himself with a foot in each camp: 'I could explain to the biologists the limitations of what the space engineers could do, and I could explain to the engineers the kind of things the biologists needed.' The process that would make Lovelock the right person to come up with the idea of Gaia was nearly complete.

The need to build lightweight but durable instruments, capable of surviving the rigours of a space voyage and landing on Mars, then analysing the soil while using no more power than a pocket torch, before sending data back to Earth using about as much power as a household light bulb, appealed to Lovelock far more than the idea of ordering equipment off the shelf to run in a sleek, air-conditioned lab in Houston: 'I learned an important lesson in those days. The atmosphere at JPL reminded me of my work during the Second World War. Think of radar. Engineers had to make tough, well-engineered equipment that would work in an aircraft flying over hostile territory.

And they had to make it quick. The situation at JPL in the sixties, with the race against the Russians, was like that. It showed what can be done if there is a sense of urgency, a mission and a will to succeed. And that's what we need today to combat global warming. The sense of purpose we had during the war, or during the space race.' It was also done, though it seems almost unbelievable now even to people old enough to remember those days, without personal computers or even electronic calculators. The few computers that existed were huge, expensive machines filling air-conditioned rooms and tended by teams of operators, but literally no more powerful than the chip in your washing machine today. Only a little more than forty years ago, spacecraft were designed and their trajectories (including soft landings) calculated using pen and paper, the power of the human brain, and slide rules.

But it wasn't all scientific calculation and experimentation. Lovelock also learned about business while he was in the United States. With his friend Al Zlatkis, he formed a company called Ionics Research, through which they could sell their skills to industry. As an individual, if Lovelock had been a consultant under contract to one firm he would probably have had to work for them on an exclusive basis; but as a company it was easy to work for different businesses at the same time. After he returned to England, Lovelock resigned from Ionics Research and formed his own company, Brazzos Limited, to carry on the good work. The name was chosen as a deliberate mis-spelling of that of the Brazos River in Texas. What Lovelock hadn't realized is that in Spanish 'brazzo' means arm; because of his associations with the Ministry of Defence, it was assumed in some quarters that Brazzos was an arms company, and it got on the list of approved NATO suppliers.

The return home came in 1963, when Lovelock rejected the riches of a $40,000 a year lifestyle in the world's most affluent country for the pleasures of rural life in an English village – not least, the health service ('There was always a nagging fear that in the States you could be ruined financially by a severe illness') and complete independence as a freelance scientist and consultant. Another factor in the move was Helen's increasing unease in Houston. Although the couple did not realize it at the time, she was suffering from the early stages of multiple sclerosis; naturally, the less well she felt the more she wanted to be back home in familiar surroundings. Once again, though, Lovelock hesitated about making the leap to total independence.

To test the water, he applied for a job as Director of the MRC's Radiation Laboratory, at Harwell. He was called for an interview in July, which conveniently meant that it could be combined with a family holiday in England. Equally conveniently, and as Lovelock had hoped, he was turned down for the job, but as he had anticipated the news that he was planning to return to England had got out, and within a week he was offered a post running the Biophysics Division at Mill Hill when the present head retired in 1964. He turned it down as tactfully as he could: 'I could never have been an administrator, and I explained to them that I wanted to work as an independent scientist.' Sir Charles Harington, still the Director at Mill Hill, not only knew Lovelock well enough to accept the decision, but passed on the news to Lord Rothschild, of Shell. Rothschild wasn't just a businessman – he was also a highly respected biologist and a Fellow of the Royal Society. He agreed with Harington that Lovelock was a special case, a man who worked best on his own. The result was an offer from Rothschild for Lovelock to work as a consultant for Shell, at an annual fee of £1,500, based at home but visiting the Shell Centre in London about once a month. 'With my $6,000 a year from JPL for three or four visits a year, I now had just enough. It was a drop in income of about 75 per cent, but that thought barely crossed my mind.'[1]

Back in America, Lovelock resigned his post at Baylor with effect from December 1963, and managed to fit in a visit to Shell's American research centre at Wood River before the family returned home in time for Christmas, settling in a rented house in Bowerchalke called 'The Laurels'. The visit to Wood River gave him plenty to think about as he turned his attention to the problems of a scientific world quite different from that of pure biochemical research. One of the

1. It also meant much longer commuter flights to JPL. 'I learned very quickly that travelling tourist class was false economy. I wasted so much time getting over jet lag that it wasn't cost-effective. Ever since, wherever I go I've always travelled First Class, and arrive refreshed and ready to set to work.'

surprises he encountered was the long timescale on which big companies plan. Governments tend to look ahead only as far as the next election, if that; but a big business like Shell plans twenty years or more into the future, and in Lovelock's experience businesses have been, at least until very recently, much more concerned about damage to the environment than governments – for the hard-headed business reason that if environmental degradation occurs to the point where there is economic collapse they will lose a lot of customers. 'Self-styled environmentalists who see all big companies as evil despoilers of nature are very far from the mark. Shell made pesticides like dieldrin and DDT in response to a public demand, to save lives and provide more food, as well as to make a profit. But as soon as the problems with these pesticides were appreciated, Shell stopped making them – long before governments acted and it became illegal to carry on. Today, it's clear that companies like Shell will in future make bigger profits by making products that alleviate pollution and global warming, so they are bound to follow that course of action.'

Financially, things turned out better than Lovelock could have dreamed. As well as his work for Shell and JPL, he soon took on a consultancy for Pye bringing in £3,000 a year, and there were other 'bits and pieces'. In 1964, his income was £10,000, which he estimates as being equivalent to about £500,000 today. Over the next few decades 'money came in like water flowing over a waterfall. It felt like if you needed anything you left a cup out near the waterfall, knowing the spray would fill it; but if you got greedy and tried to dam it, all Hell would break loose.' It was just as well. The Lovelocks' dream when they returned to England was to have a house built for them in Bowerchalke, and in fulfilling that dream they fell into the hands of an unscrupulous architect who was in cahoots with an equally unscrupulous local builder. A project initially estimated at £4,000–£5,000 eventually came in at £13,000, and 'it wasn't even finished off properly; it leaked and we had to spend more money putting that right.' But Jim says that he 'didn't really feel bad about it, it was just part of life'. Jim and Helen never gave the house, completed in September 1964, a name; but when he visited the village many years after moving on to Coombe Mill, he was tickled to find that it had become known as 'Lovelocks'.

Jim's relationship with Shell was the bedrock of his life as an independent scientist for decades, and he says that they were the only people he worked with who actively supported his work on Gaia. But he was always careful not to have all his eggs in one basket, reckoning that to be truly independent a freelance needs at least three 'employers' at any one time, so that it is always possible to choose projects to work on on merit, not because of a need for money. Given his genuinely independent turn of mind, it is perhaps not as surprising as it might otherwise seem that the father of Gaia theory and guru of the Greens should not only have close contacts with big business but also have a long working relationship with the security services.

It began in 1965, when a friend at Yale asked Lovelock if it would be possible to find people hiding in dense tropical forest using one of Lovelock's sensitive detectors. In essence, do people emit a distinctive smell that made it possible to detect them using devices like the ECD? It was pretty obvious what was behind the question, although the word 'Vietnam' wasn't mentioned, but a quick calculation showed that the idea wouldn't work – natural chemicals released by the body are not suitable for detection by the ECD. It set Lovelock thinking, though, and he realized that the trick would work if the people had been 'labelled' with a trace of a chemical that the ECD was sensitive to. It would then be easy to find them even if they were hiding hundreds of yards away. Having learned some business sense while he had been in the USA, the next time Lovelock visited JPL he had his idea notarized, establishing his right to pursue any patentable developments, and mentioned it to two of his colleagues, Dian Hitchcock and Gordon Thomas. Through their contacts in Washington, he was introduced to the CIA, where to his delight a meeting was arranged in what turned out to be an office above an antique shop. 'It was like something out of *The Man From U.N.C.L.E.* You went through what seemed to be an ordinary shop, then through an ordinary door at the back, then upstairs to the CIA office.' But neither the CIA nor the American military took Lovelock's proposal seriously. 'I think I came across as a mad foreign scientist trying to scam them with a bogus invention.'

Back in England, Lovelock began to think that it was just as well the Americans had turned him down. As a loyal citizen, perhaps he should have offered his idea to the British authorities first. He mentioned his concerns to Lord Rothschild, an establishment figure who had fingers in many pies. The old boy network swung into action, with the result that Lovelock was taken seriously and was requested to develop such detectors for use in surveillance work. The most puzzling aspect of the British approach, he found, was their concern that the chemicals being used must be harmless for the people being labelled. 'I hadn't expected that MI6 would care much about the health of KGB agents. It wasn't at all like James Bond.'[1] But he was able to reassure them that the compounds he planned to use, perfluorocarbons, are utterly safe – if anything, safer than water. 'You can drown in water,' Lovelock says with a smile, 'but perfluorocarbons carry oxygen almost as well as air. People have had their lungs filled with the stuff and experienced nothing more than discomfort.' The Secret Service were delighted.

After the necessary security checks ('It was just as well I'd joined the Catholic Society, not the Communists, back in my student days'), Lovelock was given more or less *carte blanche* to set up a group working on chemical tracers at the Admiralty Materials Research Establishment at Holton Heath, in Dorset. Until then, his expenses and fees for consultation visits had been paid by cheques drawn on Coutts, the Queen's bankers; as a consultant at Holton Heath he got what amounted to a salary. 'The work was so secret that there was no bureaucracy – nobody was allowed to know what we were doing, so there was no form-filling or administration meetings. It was wonderful!' But although what they were doing with chemical tracer technology was a secret, it was decided early on that the technology itself was not. So when Lovelock received a request for advice from the US National Oceanic and Atmospheric Administration (NOAA) for help in a project to study the way air masses move over the continental United States, he was able to provide the assistance they needed; the reason for NOAA's interest was that they wanted to know what would happen to the fallout from a nuclear accident or

1. Lovelock's own taste in spy stories runs more to Len Deighton and Gerald Seymour. Seymour, in particular, he recommends as giving what seems from his experience to be an accurate portrayal of this kind of work.

a release of toxic chemicals.[1] And when Shell needed a technique to 'label' the gases in their pipelines so that detectors would be able to tell when the flow changed from one gas to another, he was able to oblige. Everything fitted together beautifully, and once the Holton Heath lab was established Lovelock reverted to his preferred role as an adviser. 'Somehow, all the jobs I did for different companies and different government agencies fed into one another and into Gaia.' The NOAA contract is a good example. It brought in $100,000, the biggest single payment Lovelock ever received. After tax, most of this went on the latest state-of-the-art Hewlett Packard desktop computer, which cost him £20,000 in the late 1970s. It was used to write all his early books on Gaia, as well as for his scientific work. 'Life is short,' says Lovelock, 'and if you get some extra money there's no point in hoarding it.' But some lives are longer than others. The relationship with the military continues to the present day; when we visited Lovelock shortly after his 87th birthday he gleefully informed us that he had just signed a contract extending his consultancy with the Ministry of Defence for another year – 'With luck they'll keep me on until I'm 90!'

This long relationship with our guardians has left Lovelock with a much more rosy view of the security services than most people have. 'The ones I've met remind me of the friendly bobby on the beat that I knew as a child in Brixton. One of them was even a Quaker. My strong impression is that we are in good hands – but of course, I don't know what goes on at the operational end of things.' The secret services are, though, secretive, and this rubbed off on Lovelock, who learned the value of discretion. It was a great asset, he realized, for a freelance consultant who worked for different companies and inevitably picked up information that would be useful to rivals. His other major employer during the last four decades of the twentieth century, Hewlett Packard, is an archetypal example.

The relationship with Hewlett Packard went back to the early 1960s, when Lovelock's company Ionics Research worked with a small American outfit called F and M Scientific, in the business of

1. Years later, the American military did start using the technique, having learned about its civilian applications.

manufacturing gas chromatographs. Ionics Research gave way, as far as Lovelock was concerned, to Brazzos, and F and M got taken over by Hewlett Packard, but Lovelock maintained a relationship with the organization. One of the advantages was that from the mid-1960s onwards he always had the latest computer technology, starting with electronic calculators and progressing to the best desktop PCs, to carry out his work. These computers made it possible for him to overcome his algebraic dyslexia, getting the machines to do the calculations that he found so confusing: 'without them, I could never have developed Gaia into a proper scientific theory.' The work for Hewlett Packard involved chromatography, detector design, and simply coming up with bright ideas which they could patent. When he worried that, as in the Mill Hill days, he might be producing too many ideas which would never see the light of day, he was reassured by a Hewlett Packard executive who told him that the company employed too many lawyers, and it was best to keep them busy. If they were working on patent applications, if nothing else it would keep them out of mischief. The truth, of course, is that it is fine for a company like Hewlett Packard, sifting through and patenting a hundred or more ideas, if just one of them turns out to be the next ECD. The value of Lovelock's consultancy to the company was clear financially – 'by 1994, when I stopped working for them, my annual income from Hewlett Packard had reached $32,000.' But Lovelock only ever patented one of his inventions himself. 'It always seemed too much bother, and we had enough money coming in. I suppose if we'd been hard up I might have tried to exploit my ideas commercially, but there were other things to think about.' Things like Gaia. By the early 1960s, the ECD was beginning to affect the way people thought about our planet, even though the Gaia idea had not yet emerged; over the following decades it would be instrumental in exposing the biggest human impact on the atmosphere up to that time. This simple, hand-held device would revolutionize our understanding of the relationship between human activities and the environment, and help to kick start the Green movement.

6

Green Revolutions

Rachel Carson's book *Silent Spring*, first published in 1962, is widely regarded as responsible for the initial impetus given to the 'green' movement – the book that blew the whistle on humankind's detrimental impact on the environment. It certainly had a big influence, both on the general public and on policymakers and the chemical industry; but as with so many influential books the interpretation put on the author's work is not always what the author seems to have intended.

Carson was born on 27 May 1907, on a small family farm near Springdale, Pennsylvania. Her interest in the natural world was stimulated by her mother, who taught her about the life inhabiting the fields, forests and ponds nearby; although she started out studying English and creative writing, she soon switched to marine biology, and ended up combining both interests in her work. She graduated from what was then the Pennsylvania College for Women (now Chatham University) in 1929, and started studying for a PhD at Johns Hopkins University, attending the Woods Hole Marine Biological Laboratory every summer and supporting herself by teaching, working in the zoology department of the University of Maryland. But her already precarious financial situation got worse in 1932, when her father died, so she had to give up her hopes of a PhD, converting the work she had done so far into a Masters thesis, and working to support her elderly mother. She was employed by the US Bureau of Fisheries in Washington, DC, initially as a scriptwriter for radio shows, then in 1936 passed the Civil Service exam and obtained a job with them – only the second woman hired by the Civil Service in a full-time, professional post – working on radio scripts, books and other publications. She also wrote newspaper articles to bring in a little extra money.

In 1937, Carson had an article on life beneath the surface of the sea published in *Atlantic Monthly*, and the same year her responsibilities increased when her elder sister died at the age of 40, leaving Rachel to care for two young nieces. The *Atlantic Monthly* article was eventually developed into a book, *Under the Sea-Wind*, which was unluckily published a month before the Japanese attack on Pearl Harbor and, hardly surprisingly, failed to find an audience in a country preoccupied with war. By 1949, Carson was Chief Editor of Publications of the Fish and Wildlife Service, as the Bureau of Fisheries had become, and had been working for years on a second book, using a wealth of material that had been classified during the war years. It was eventually published as *The Sea Around Us* in 1951. It became a smash hit, staying on the *New York Times* bestseller list for eighty-six weeks and making Carson financially independent by the time she was 45. She gave up her job in 1952 and published *The Edge of the Sea* in 1955; it achieved comfortable success, though nothing on the scale of *The Sea Around Us*. Her life changed again in 1956 when one of her nieces died at the age of 36, leaving a five-year-old son, Roger Christie; Carson adopted the boy and bought a property at Silver Spring, in Maryland, in order to raise him in a rural environment – she was also still caring for her mother, now nearly 90, who survived until 1958. Hardly surprisingly, in the light of these commitments, Carson herself never married; she died, of breast cancer, on 14 April 1964, not long after the publication of *Silent Spring*.

The fact that Carson knew she was seriously ill with cancer at the time she was writing her seminal book may explain some of the over-the-top passages in it, which did a lot of good in making people aware of the need for caution when interfering with the 'balance of nature', but which don't pass muster as a cool scientific appraisal of the problem. What Carson specialized in was passionate, poetic writing that got at her readers emotionally, as in the passage which gave her the title of the book:

> It was a spring without voices. On the mornings that had once throbbed with the dawn chorus of robins,[1] catbirds, doves, jays, wrens, and scores of other

1. The American robin – not the different, but similarly coloured, European bird of the same name.

bird voices there was now no sound; only silence lay over the fields and woods and marsh.

This is, of course, fiction – it is Carson's vision of the potential future if chemical pesticides continued to be used indiscriminately, not a report of anything Carson had actually seen. But it was based on a nugget of fact. In January 1958, soon after she had moved to Maryland, Carson heard from a friend, Olga Huckins, who owned a bird sanctuary. The sanctuary, like all its surroundings, had been liberally sprayed from the air with insecticide, as part of a government programme to eradicate pests. Many of the birds were directly affected, while many beneficial insects died along with the pests; her friend asked Carson, by then a famous biologist, if she could exert any influence on the government to investigate pesticide use and control it more effectively. Carson had for some time been worried about the widespread use of pesticides; following this letter, the more she looked into the matter, the more concerned she became, and the upshot was that she spent almost four years researching and writing what became *Silent Spring*.

The background to the book was the widespread use of insecticides, in particular DDT, in the post Second World War era. The power of DDT (dichloro-diphenyl-trichloroethane) as an insecticide had been discovered in Switzerland in the late 1930s, by Paul Müller, a chemist working for the Geigy company. It was developed on an industrial scale in the United States during the war and used on a massive scale to combat typhus (spread by lice) and malaria (spread by mosquitoes), which were causing crippling losses to American troops – according to General Douglas MacArthur, before the introduction of DDT at any one time two thirds of the troops in the South Pacific were afflicted with malaria. Under wartime conditions, vast areas were covered with DDT using bombers carrying containers of the insecticide under their wings; in a famous example of the power of DDT, in 1944 a potentially catastrophic typhus epidemic in Naples was stopped in its tracks by dowsing a million people with DDT powder.

There is no doubt that DDT and other insecticides did a lot of good – and still do a lot of good. In Italy, there were 411,062 cases of malaria in 1945, and only 37 in 1968. In the southern United States, malaria has been essentially wiped out, and many people who live there don't even know that the disease used to be endemic. Tens of millions of lives have been saved by DDT alone in countries such as India and the Philippines. The trouble was that the wartime approach, the 'can do' mentality that saw mankind as supreme over nature, carried over, particularly in the United States, into peacetime. To take just one example, when Dutch Elm disease (a problem caused by beetles in the bark of the trees) affected the campus of Michigan State University in the mid 1950s, the whole area was liberally sprayed with the poison. In an unanticipated consequence, leaves coated with poison fell to the ground and were eaten by earthworms who were in turn eaten by robins, who died. Or as Carson put it, in her lyrical way,

> we spray our elms and following springs are silent of robin song, not because we sprayed the robins directly but because the poison traveled, step by step, through the now familiar elm leaf–earthworm–robin cycle. These are matters of record, observable, part of the visible world around us. They reflect the web of life – or death – that scientists know as ecology.

There is another aspect to the problem, also discussed by Carson. Pests become resistant to pesticides. When a new pesticide is first introduced, it kills a lot of pests, but the individuals who survive are the ones with the strongest natural resistance to the pesticide. When they breed, their offspring inherit this resistance. Over many generations – and insects breed quickly – this builds up a large population of resistant individuals. Like the development of strains of disease organisms resistant to drugs ('superbugs'), it's an example of evolution at work, or in Carson's words, 'a triumphant vindication of Darwin's principle of the survival of the fittest'. The more miraculous a wonder drug or pesticide seems to be, the more important it is to use it thoughtfully, holding it in reserve for cases of real need, so that it will retain its effectiveness as long as possible.

In her discussion of this, and her examples of the direct threat to the web of life, Carson was on solid ground. She did not say that all chemical insecticides should be banned, but that they should be used with discrimination to target specific pests. She also urged the use of biological controls, introducing natural predators to eat the pests – although it has to be said that this approach also requires extreme caution lest the cure turns out to be worse than the disease. But she did also exaggerate the problems, especially in her chapter concentrating on the alleged cancer-causing potential of chemical pesticides.

In publicity terms, this was a masterstroke. In the early 1960s, even more than today, cancer was the big health fear, the unspeakable disease. It's one thing to be concerned that pesticides may be killing robins; quite another to be worried that they might be killing you. Among the 'evidence' Carson included in her book was the example of a woman who sprayed her basement with DDT to eradicate an infestation of spiders and became ill. 'When examined by Dr [Malcolm] Hargraves she was found to be suffering from acute leukaemia. She died within the following month.' And there is also the example of another patient of Dr Hargraves, a man who sprayed the basement of his office building to get rid of cockroaches. 'Within a short time he began to bruise and bleed. He entered the clinic bleeding from a number of haemorrhages. Studies of his blood revealed a severe depression of the bone marrow called aplastic anaemia . . . nine years later a fatal leukaemia developed.' Such anecdotal examples are laughable as 'evidence' that DDT causes leukaemia. But they are still in the current edition of the book, and still causing concern. For that reason, it is worth stressing that four decades after *Silent Spring* was published, in spite of many properly conducted studies, no evidence at all has been found that DDT causes cancer in people. The cancer chapter is the worst blot on what is, overall, still an important book, if Carson's message is correctly interpreted.

The book first appeared as a serialization in the *New Yorker* in June 1962, and between hard covers later that year. It provoked a furious response from many sections of the chemical industry, but the resulting publicity only served to boost sales; the Book of the Month Club alone printed 150,000 copies just for its initial mailing to members. Above all, whatever its faults, *Silent Spring* boosted public and government awareness of the impact of human activity on the environment. The immediate response of President John F. Kennedy was to request a report from his Science Advisory Committee, which in 1963 echoed Carson's main (and now totally uncontroversial) message, pointing out that pesticides were needed to maintain the quality of Americans' food and health, but warning against their indiscriminate use, saying that uncontrolled use of such poisonous chemicals was potentially 'a much greater hazard' than radioactive fallout.

Groups such as the Natural Resources Defense Council, the Wilderness Society, and the Environmental Defense Fund were established during the years following publication of the book; the establishment of the Environmental Protection Agency in 1970 owed much to Carson's legacy; and DDT was banned entirely in the United States in 1972. As Al Gore wrote in an introduction to the 1994 edition of *Silent Spring*, the book 'brought environmental issues to the attention not just of industry and government; it brought them to the public'. Without this book, he said, 'the environmental movement might have been long delayed or never have developed at all.'

But why did responsible bodies such as the President's Science Advisory Committee take the pesticide threat so seriously? It wasn't just because Rachel Carson wrote so powerfully and tugged at their heart strings. It was because exactly at the time her book was alerting the public to the dangers in emotional terms, hard scientific evidence was coming in that pesticides such as DDT and dieldrin were far more ubiquitous in the environment than anyone had realized.

Just at the time Carson was writing her book, the electron capture detector was being used independently by scientists at Shell's research centre in Kent and at the US Food and Drug Administration to measure the residue of pesticides in the environment. It quickly became clear that pesticides were present everywhere, including in the tissues of people and other animals. In Lovelock's words, 'this lent veracity to Carson's otherwise unprovable statements.' Although she rightly gets credit for sounding the warning in clear language and getting the public emotionally involved, 'people often forget the positive role played by scientists in both industry and government agencies in establishing the reality and magnitude of the problem.' It was evidence obtained using electron capture detectors, not the emotional arguments alone, that finally led to the DDT ban in the United States.

At first, Lovelock was delighted by this use of the ECD, since even in those pre-Gaia days he shared Carson's views on ecology and the web of life, and was already concerned about damage to natural ecosystems. When she wrote 'the earth's vegetation is part of a web of life in which there are intimate and essential relations between plants and the earth, between plants and other plants, between plants and animals,' she struck a chord. In a CBS documentary in which she appeared in 1964, not long before her death, Carson said:

> Man's attitude toward nature is today critically important simply because we have now acquired a fateful power to alter and destroy nature. But man is a part of nature, and his war against nature is inevitably a war against himself.

This is a message that resonated with Lovelock at the time, and resonates with us today. But, says Lovelock, 'you have to have a bit of common sense' when using an instrument as sensitive as the ECD. It is so sensitive that 'it can detect absolutely trivial quantities of pesticides and other pollutants. Even organically grown vegetables contain measurable amounts of pesticides, if you use an ECD to make the measurements.' This means that it doesn't make sense to set the lowest permissible level of pesticides in food, for example, as zero, because 'you'd have to reject nearly everything we eat.' Lovelock likes to quote the sixteenth-century German physician Paracelsus, who said, 'the poison is the dose,' and points out that pure water will kill you if you drink too much of it, while the deadliest nerve gases are not deadly at all in quantities measured in a few picograms, which the ECD can easily detect. By nature a quiet man, for all his determination, one of the few things that rouses him to anger is the way 'self-styled Greens' who 'not only know nothing of science, but actively disdain science', are 'happy to use the results from ECDs and other scientific instruments to support misguided crusades to get everything banned. Even if we restrict ourselves to DDT, they forget that even today it saves millions of lives in the Third World.[1] If that means a measurable but insignificant trace of DDT getting into the food chain, it doesn't matter, and I'm sure Rachel Carson would agree.'

The sensitivity of the ECD has, however, led Lovelock to become

1. There's a particular irony about the need to use DDT and other pesticides to tackle malaria. As the world warms, the mosquito that transmits malaria can move into the highlands of affected regions, as well as spreading out from the tropics. In 2006, the World Health Organization (WHO) estimated that the spread of malaria caused by climate change is already putting an extra 20 million people at risk in Kenya alone.

aware of a phenomenon that, he feels, deserves much more attention than it gets. Early on in his work with the detector, he noticed that there is a relationship between the affinity molecules have for electrons and their biological activity. Many important compounds in the biological system of energy transport in cells have a strong affinity for electrons, which behave in many biological reactions as active particles. This is not a surprise, and can be explained in terms of the chemistry involved, though this is not the place to go into details. But Lovelock then found that many non-biological materials that are very efficient at absorbing electrons are carcinogenic. Given that anything which can absorb electrons must interfere with biological processes that absorb electrons, this is a plausible link, but one that has never been followed through with proper studies to find out what is going on. 'Whenever I come across a chemical substance that is a strong electron absorber, I treat it with caution.' When he first noticed the link, and drew it to the attention of his medical colleagues, they dismissed it, pointing out that there were well-known compounds, such as trichloroethylene, that were known electron absorbers but were so safe they were used in anaesthesia, and were not known to be carcinogenic. But '"not known to be carcinogenic" is not the same as "known not to be carcinogenic,"' says Lovelock. 'Now, we know differently. Many of these substances have since been found to cause cancer. I wish someone would look into this connection properly.'

The ECD played a supporting role in the story of *Silent Spring* and the revolutionary change in attitudes towards the environment that followed; the passion and the fear came first, and then the science backed the fear up. In its most spectacular and best-known contribution to environmental awareness, however, the ECD measurements came first, and the concern followed. This is, of course, the story of chlorofluorocarbons (CFCs) and their impact on the ozone layer. It is worth elaborating here, even though telling the whole tale means getting ahead of our story chronologically, not just because of Lovelock's personal involvement (much more personal than his involvement in the *Silent Spring* story) but because it is now regarded as a success story, an archetypal example of how appropriate and timely action can heal the damage done to the planet – to Gaia – by thoughtless human acts.

As far as Lovelock is concerned, the story began one grey day in January 1965, when his daughter Jane, visiting home on a break from her work as a trainee nurse in Southampton, announced that she had a week off and suggested a family holiday abroad. The Lovelocks being the Lovelocks (and air travel being rather less accessible in those days), they settled not for the Canary Islands or some other obvious winter hotspot, but for the west of Ireland – almost as far west as they could go, at Kenmare, in the southwest of the island.

Taking their car by sea on the fifteen-hour voyage from Swansea to Cork then driving west through the beautiful, sparsely inhabited countryside, the family fell in love with the region. They returned in the summer to spend two weeks in a rented cottage, and after the first week decided to buy a cottage that they had found on the south coast of the Beara Peninsula, near the village of Adrigole. It cost £3,000 – a little on the expensive side for the time, but as it turned out probably the most important purchase Lovelock ever made. Returning to Adrigole at least once a year throughout the 1960s, Lovelock became aware that compared with the clean air of Ireland, back home in Wiltshire the atmosphere had become increasingly hazy, spoiling the views across the countryside by limiting visibility to less than a mile, while stealing the warmth from the Sun. He was sure that what he was witnessing was some kind of summer smog, a phenomenon caused by pollution. It became something of a family ritual at Bowerchalke to measure the effect of the haze on the direct rays of the Sun, using a hand-held photometer like those used in photography. But Lovelock was disappointed to find that there were no official records which would enable him to find out whether the haziness had indeed increased during the decades of increasing industrial activity. He decided instead to test his hypothesis by looking for some sort of atmospheric 'tracer' that could only have come from urban industrial regions and had no natural sources in the countryside. The obvious choice was the CFC family, then still widely used in refrigerators and increasingly as the propellant gases in so-called aerosol sprays. CFCs had been chosen for these uses because they are very stable compounds that scarcely react with anything at all. Because they do not react, they are non-toxic and non-inflammable, which seemed to make them ideal for household use. But because they do not react they are very long lived and stay around in the atmosphere for a very long time, gradually building up their concentration in the air as more and more of them are released.

Starting in 1969, the Lovelocks began measuring the concentration of the compound CFC-11 in the air at Bowerchalke, together with simultaneous measurements of the haziness and the wind direction. As a control experiment, they also started measuring the same things at Adrigole, to get a benchmark for clean air – since the prevailing winds come from the west, across the Atlantic Ocean, the west coast of Ireland is as free from pollution as anywhere in Europe. Even at Adrigole, there were days when the wind came from the east and a smog-like haze filled the air. But there were many more days when the wind blew from the west and the air was clear.

It was no surprise to Lovelock to find CFC-11 in the air at Bowerchalke, with more CFC present when the air was more hazy. He also found about 150 parts per trillion of CFC-11 in the air even at Adrigole on hazy days, confirming that the haze was pollution blown a thousand kilometres or more westward from Britain and mainland Europe. But he was surprised to find 50 parts per trillion of CFC-11 in the air even when the wind came from the west and it was clear at Adrigole. Could it have come all the way from America? Or was the entire atmosphere of the Earth contaminated with CFCs? Lovelock's discovery that England was indeed 'plagued by the same kind of smog as Los Angeles' was published in the *Journal of Atmospheric Environment*, though it 'roused no interest at the time'. But by the time it was published, he was already off on the trail of the CFCs that seemed to be coming in off the Atlantic. How far had human pollution spread? He realized that these CFCs would be ideal tracers for finding out how human activities were affecting the atmosphere, and that the best way to trace human pollution (which mostly originates in the Northern Hemisphere) around the world would be to take an ECD on a ship voyaging to the Southern Hemisphere and see how far the CFCs had got.

Purely out of scientific curiosity, together with Robert Maggs, a colleague at Reading University, Lovelock applied to the Natural Environmental Research Council (NERC) for a small grant to make an instrument to measure CFCs, dimethyl sulphide (DMS) and methyl iodide and take it on board the research ship *Shackleton*, which made an annual voyage to the British scientific station in Antarctica. The proposal was turned down because nobody on the academic committee handling the grant applications was aware that the ECD really could measure concentrations of CFCs at the level of parts per trillion; they dismissed Lovelock as a crank trying to con his way into a free ride to Antarctica and back. Fortunately, some of the civil servants on the staff at NERC knew better, and Lovelock was offered free passage on the *Shackleton* as far as Montevideo, provided he paid for all his equipment himself. He built the apparatus in a few days in his lab at Bowerchalke, and in the event it worked perfectly throughout the six-month voyage. 'The total cost was a few hundred pounds,' he says, 'and it led to the so-called "ozone war", to major research programmes on the links between life in the sea, DMS and climate. Thousands of scientists have been kept busy for decades following up what we found on that voyage.'

The *Shackleton* was a converted Baltic coaster, displacing a few hundred tonnes, in which the scientists of the British Antarctic Survey found space to give Lovelock the use of a bit less than five metres of bench with cupboards underneath and standard 'mains' power supply. 'That was fine by me – more than I needed, in fact. My quarters were even better, in the sickbay where the doctor usually lived. We sailed in November 1971, and I started taking measurements of CFCs before we left Barry docks.'

Lovelock took his air samples at the bow of the ship, right behind one of the two holes through which the anchor chains passed, where the air blew straight in off the sea. 'The other scientists on board worked so hard I felt almost embarrassed how easy it was to make the CFC measurements.' The big change in CFC concentrations came, as expected, when the ship crossed in to the Southern Hemisphere. As far as the atmosphere is concerned, the boundary between the Northern and Southern Hemispheres doesn't lie along the geographical equator, but at a boundary called the intertropical convergence zone (ITCZ), which shifts north and south over the course of the year, in step with the seasons. The ship crossed this boundary during one night, and 'the next day it was striking how clean and unpolluted the air was. The air of the Northern Hemisphere is always a bit hazy;

1. Jim's parents, Nell and Tom (*right*), on their wedding day

2. The family shop in Brixton

3. Jim Lovelock in 1922

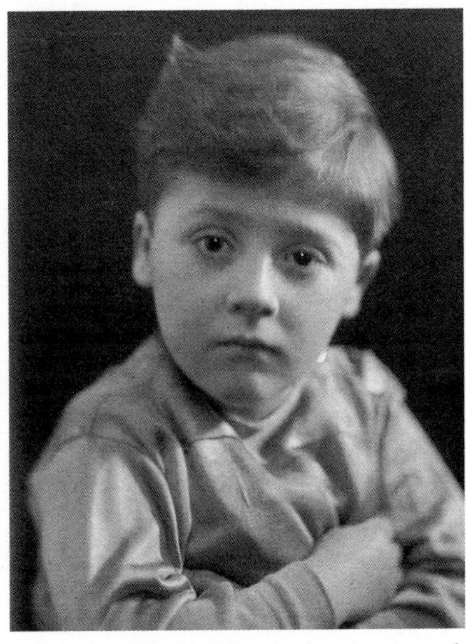

4. By 1924, Jim was already showing signs of his independent nature

5. Kit with Hugo's father, 'Papa' Leakey

6. Jim's Aunt Kit and Uncle Hugo Leakey were a big influence on Jim as an adolescent

7. Jim in his new uniform as a grammar school boy

MEDICAL RESEARCH COUNCIL

Telegrams:
'NATINMED HAVER, LONDON.'
Telephone:
HAMPSTEAD 2232.

NATIONAL INSTITUTE
FOR MEDICAL RESEARCH,
HAMPSTEAD,
LONDON, N.W. 3.

6th June, 1941

I.E. Lovelock Esq.,
41 Hillview Rd.,
Orpington, Kent.

Dear Sir,

Professor A.R. Todd has suggested to me (via Sir Henry Dale, F.R.S., the Director of this laboratory) that you might care to apply for the appointment which is outlined on the attached note. If so, would you let me know whether you could call here for interview some time in the next week. The work is not likely to involve much organic chemistry but will involve the use of varying types of physical and chemical apparatus. Advanced mathematics are not essential.

Yours truly,

R. B. Bourdillon

8. The letter inviting Jim to apply for a job with the Medical Research Council

9. Jim (*left*) with two of his wartime colleagues in their unmarked battle dress; they dubbed themselves 'the three musketeers'

10. HMS *Vengeance* in Arctic waters

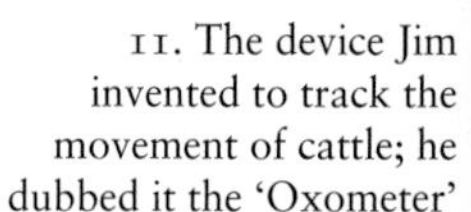

11. The device Jim invented to track the movement of cattle; he dubbed it the 'Oxometer'

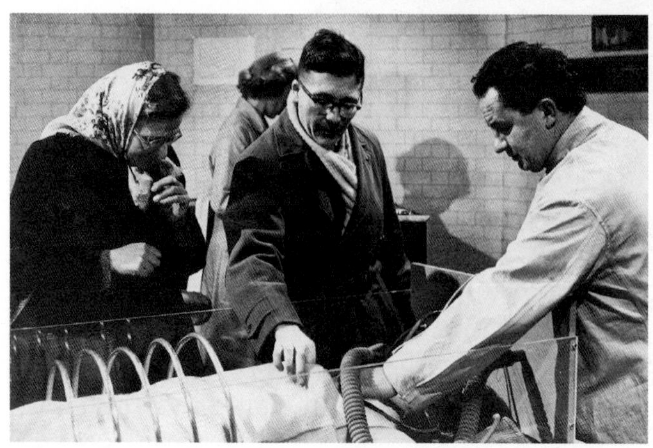

12. Jim on the set of the TV drama *The Critical Point*, offering technical advice

13. Jim's laboratory in Bowerchalke

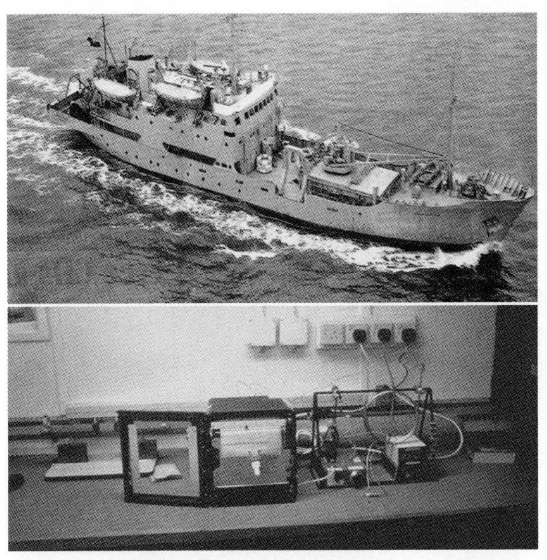

14. The RV *Shackleton* and the tiny bench space where Lovelock worked on board ship

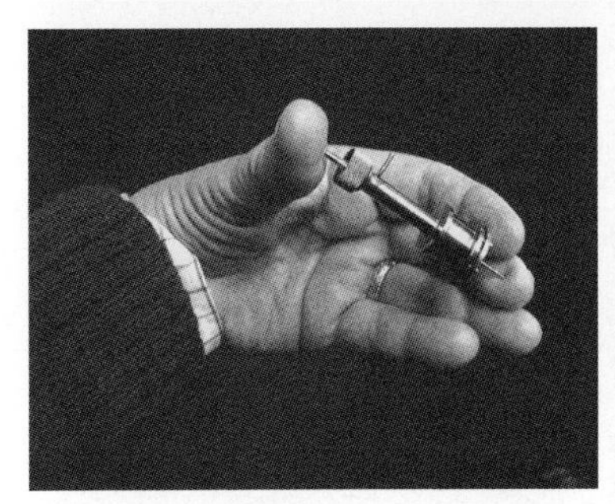

15. An electron capture device (ECD) in Jim's hand

16. Jim collecting air samples from the bow of the *Shackleton*

17. Jim returning to the *Meteor* after taking air samples in a rubber dinghy far away from the polluted air on board the ship

18. *Above*: The lab at Coombe Mill

19. *Right*: Jim in the lab

20. The Lovelocks on their visit to the French nuclear processing plant at Le Havre

21. Part of the coastal path walked by the Lovelocks

22. Jim contemplating the route along the coastal path

23. Jim standing by a statue of Gaia in the garden at Coombe Mill

the southern air was so clear that even though we had been working on deck ever since leaving Britain, people suffered from sunburn.' In crossing from the atmospheric Northern Hemisphere into the Southern, the concentration of CFC-11 in the air fell from 70 parts per million to 40 parts per million – but it didn't fall to zero. This was already a significant discovery, alone justifying Lovelock's presence on board the *Shackleton*.

Although Lovelock had to leave the ship at Montevideo to return home, by then he had shown one of the other scientists on board, Roger Wade, how to make the daily measurements, which continued for the rest of the voyage to Antarctica and back. These showed that there were CFCs present throughout the Southern Hemisphere. Since CFCs are only produced by industrial activity, this meant that pollution from human activities had spread completely around the world. When Lovelock wrote the discovery up for a paper in *Nature*, he wanted to emphasize that there were no medical risks associated with CFCs. 'I've never been a doomwatch sort of person,' he says, 'so I wrote that they presented "no conceivable hazard".' He shakes his head at the memory. 'I should have said, no conceivable *toxic* hazard.' Within a couple of years, the discovery of CFCs in the air over the Antarctic Ocean was being used as ammunition in what became known as the ozone war.

Ozone is a pale blue gas that is poisonous to human life even in small concentrations. It is a form of oxygen with three atoms in every molecule (O_3) instead of the more common two (O_2), and can be produced at ground level by reactions involving sunlight acting on polluting chemicals; ozone at ground level is a component of photochemical smog. But at a safe distance, ozone is essential to our wellbeing. Ozone in the region of the atmosphere known as the stratosphere protects the surface of the Earth by blocking ultraviolet radiation from the Sun which might otherwise make life difficult over most of the land area of our planet. But while ultraviolet radiation might not be good for life forms adapted to high-latitude conditions today, Lovelock is not convinced that it would do much harm to Gaia. 'The UV in Kenya,' he points out, 'is eight times that in the UK. But you don't find any sunburned trees or lions in Kenya.'

The atmosphere is most simply described as a series of layers, defined in terms of the way temperature changes with altitude. The lowest layer of the atmosphere is the troposphere, and air close to the ground (or sea) is warm because the surface of the Earth is warmed by the Sun. The warm surface of the Earth radiates heat at infrared wavelengths and warms the troposphere through the greenhouse effect. The further away from the surface, the colder it gets, until at an altitude of about 15 km the temperature is about -65°C. This is the top of the troposphere, a rather vague boundary called the tropopause. Above the tropopause, temperature increases with altitude, up to about 50 km. This warming layer of the atmosphere is called the stratosphere, or sometimes the ozone layer. It is warm because it absorbs ultraviolet energy from the Sun, preventing it reaching the ground. Ozone is constantly being manufactured and constantly being destroyed by reactions involving sunlight, but overall the amount in the stratosphere stays roughly constant, like the level of water in a bathtub where water is flowing in to the tub from the taps but flowing out through the plughole as fast as it flows in.

At its warmest, parts of the stratosphere may be as warm as -20°C, but it can get much colder than -65°C, especially above the polar regions during the long polar winter when the Sun never rises. The top of the stratosphere is called the stratopause; above that there are other layers of the atmosphere which need not concern us here. Although the stratosphere is thick in the sense that it spans a depth of some 35 km (from 15 km altitude to 50 km altitude), it is thin in the sense that the density of the gas in the layer is very low. If all the ozone in the stratosphere were somehow brought down to sea level and put under the same pressure as the air that we breath, it would form a layer only about three millimetres thick – roughly the thickness of a pound coin. There is no more than five billion tonnes of ozone in the stratosphere, a small amount of gas considering its essential role in making the Earth suitable for life. Because of this, before Lovelock's discovery of the way CFCs had spread around the globe, by the early 1970s some scientists and members of the burgeoning Green movement were already concerned that human activities might be able to damage what is a relatively frail component of the Earth System.

Their initial concern focused on the potential threat to the ozone layer posed by fleets of high-flying supersonic transport aircraft (SSTs) such as Concorde, which would come in to transatlantic service in 1976. The exhaust from such aircraft releases compounds of nitrogen and oxygen that would definitely react with ozone in the stratosphere, and this stimulated a heated debate for a couple of years, with calculations suggesting that a fleet of 500 SSTs flying for two years might cause a 10 per cent reduction in the amount of ozone in the stratosphere, but with nobody quite sure what harm that would do to the Earth System. The arguments became academic when no such fleets of SSTs were ever built, and even Concorde has now been grounded. (Other airliners fly at much lower altitudes.) But while that threat to the ozone layer faded, another grew.

In 1971, some months before the *Shackleton* voyage, Lovelock attended a conference in the United States where he fell into conversation with Lester Machta, who worked at the National Oceanic and Atmospheric Administration, and Ray McCarthy, of Dupont – the company which made most of the CFCs used up to that time, marketed under the name Freons. Lovelock told them about his early work with the CFC sniffer, and asked McCarthy how much Freon had been released since they had gone into production. His estimate turned out to be very close to Lovelock's estimate, based on ECD measurements, of how much CFC there was in the atmosphere at that time. The implication was that none was being destroyed. In January 1972, at a meeting on atmospheric chemistry held in Florida, Machta mentioned this interesting tit-bit of information to Sherry Rowland, a scientist based at the University of California, Irvine. Rowland was intrigued. If nothing was destroying CFCs in the troposphere, that meant they would spread up through the atmosphere into the stratosphere. There, ultraviolet radiation will break up any large molecule. 'Of course,' he said to Machta, 'it will always decompose with ultraviolet.'[1] But neither of them followed the remark through to its logical conclusion for more than a year.

By the summer of 1973, Lovelock's data from the *Shackleton* voyage had been published and begun to rouse interest. When a young researcher called Mario Molina joined Rowland at Irvine, Rowland encouraged him to find out what would happen to the CFCs spreading

1. Quoted by Dotto and Schiff.

through the troposphere once they worked their way up into the ozone layer, by working through all the appropriate chemical reactions. The answer – obvious once somebody thought to ask the question – was that the big molecules would be broken up by ultraviolet radiation, releasing atoms of chlorine. And chlorine reacts like crazy with ozone, acting as a catalyst which converts ozone into the diatomic form of oxygen, O_2. According to Rowland, a single atom of chlorine released into the stratosphere will destroy 100,000 molecules of ozone before itself being locked up in some less reactive form. Unlike ozone, diatomic oxygen offers very little protection from ultraviolet radiation.

Intrigued by the work of Rowland and Molina, in 1974 Lovelock arranged through his contacts at the Ministry of Defence to fly with his CFC sniffer as an unofficial passenger on a routine Hercules flight from RAF Lyneham, forty miles north of his home in Wiltshire. His measurements, published later the same year, showed a steady concentration of CFCs in the troposphere, but a decline in the stratosphere, exactly in line with the theory that they were being broken down in the upper atmosphere.

Other researchers were carrying out similar studies of CFCs at about the same time as Rowland and Molina, and by 1975 the bandwagon was rolling. Some chemists and environmental groups perceived that CFCs from spray cans and refrigerators could destroy the ozone layer, allowing ultraviolet radiation to penetrate to the surface of the Earth, posing a severe danger to life, including human life. On the other side of the battle, industry fought back, arguing that these ideas were unproven and that no action should be taken on the basis of a few speculations to curb a billion-dollar industry that employed thousands of people.

In Congressional hearings held to look into the matter, and run like a trial, Lovelock found himself in the odd position of being the chief scientific 'witness' called on behalf of Dupont. Why did he speak on their side of the debate and not on the environmentalists' side? 'Because they asked me first. If the other side had asked me, I'd have appeared on their behalf. All I did was present the scientific facts, which don't take sides. That's the job of lawyers and politicians.' It still rankles with him that as a result he was accused in some quarters of being a bought poodle of the aerosol spray industry, and he stresses that he received no payment for the testimony, just his travelling expenses. 'The adversarial approach may work in a court of law, but it is no way to settle scientific issues.'

Even outside the Congressional hearings, there were aspects of the CFC–ozone debate that infuriated Lovelock because 'the science was so shoddy.' Early on in the discussions, at a meeting in Snowmass, Colorado, he was asked to estimate the accuracy of his measurements. Any well-trained scientist knows that the accuracy of your measurements is not the same as the precision of the instruments used. A metre rule is usually calibrated in divisions as small as one millimetre, so you can measure things as short as one millimetre. But if you try to measure the length of a football pitch using a metre rule, your measurement will not be accurate to one millimetre, because each time you move the rule along to a new position you get a small error. Lovelock knew that there were small errors in measuring the amount of air taken in for each sample analysis, errors reading off measurements from the apparatus he used, and so on. Cautiously, he estimated his accuracy as plus or minus 20 per cent. This means that if you are measuring something that has a value of 100, you might get a single answer anywhere from 80 to 120. But if you make a hundred measurements of the same thing, the average of all your answers ought to be close to 100. At the meeting, American researchers claimed that they could measure CFC concentrations to an accuracy of 1 per cent. Lovelock was impressed, and felt that as a mere amateur he wasn't in the same league as the professionals. It was only years later that he discovered they meant a *precision* of 1 per cent – the equivalent of measuring a football field with a metre rule and claiming an accuracy of one millimetre. 'It still makes me mad,' he says, 'not because I was made to look an idiot but because it shows how poorly practical science is taught these days.' And he mentions another occasion when his measurements were dismissed because 'they did not fit the model'.

All the measurements were plagued by the problem of calibrating the instruments – making samples of air in which there were traces of CFCs at the level of parts per trillion to test the instruments on. It is now clear that Lovelock was the only person at the time capable of measuring these concentrations accurately, thanks to his experience of working with ECD devices that he had built himself. The way some of his claims were pushed aside still rankles after all these years, even though it is his figures that are now accepted and used in the historical data records. The accuracy of the original *Shackleton* measurements has even been confirmed by analysis of bubbles of air from the 1960s and 1970s trapped in the Antarctic ice sheets. In the end, the evidence that CFCs were likely to cause ozone depletion was overwhelming.

Whatever the value of the method used to reach the conclusion, the upshot was that, after further government-sponsored research, by the end of the 1970s the use of CFCs was severely restricted by US legislation, and in 1987 international agreement to reduce CFC emissions was reached under the so-called Montreal Protocol. But a major reason for the success of the environmental lobby in getting this international agreement was that the threat to the ozone layer was no longer just theoretical or speculative. Two years earlier, a team from the British Antarctic Survey had reported the discovery of ozone depletion over Antarctica so severe that it was dubbed 'the hole in the sky'; by 1987, there was proof positive that it was caused by the presence of CFCs in the stratosphere.

Although Lovelock was not involved in the discovery of the hole in the sky over Antarctica, the way the hole was discovered is a classic example of the kind of basic scientific research using simple instruments that has always appealed to him. Ever since 1957, the International Geophysical Year, researchers from the British Antarctic Survey had been taking measurements of the amount of ozone overhead from the same site, at Halley Bay, during the summer months from October to March. They used an instrument called a Dobson spectrophotometer, which simply looked up through the atmosphere and measured the amount of ultraviolet light being absorbed by the ozone – the depth of the 'gap' in the solar spectrum corresponding to ozone absorption. The measurements were made in 'Dobson units', and from 1957 to the middle of the 1970s every Antarctic spring (October), the instruments recorded 300 Dobson units of ozone, corresponding to a thickness of 3 mm at sea level, overhead. Then, the readings began to show a decline. By 1982, the Halley Bay instruments suggested that there was 20 per cent less ozone over Halley Bay in October than there had been for the first thirty years of observations.

The BAS team were particularly puzzled by this because they knew that NASA had launched a satellite called Nimbus 7, carrying an instrument known as TOMS (Total Ozone Mapping Spectrometer) and another called SBUV (Solar Backscatter Ultraviolet), in 1978. These instruments were designed to measure the concentration of ozone in the stratosphere – but there had been no reports from NASA of any decline. Perhaps something was wrong with the Dobson spectrophotometer at Halley Bay. Cautiously, the team, headed by Joe Farman, waited until 1984, when a new instrument, carefully calibrated back in England, started operating at Halley Bay. It gave the same result – only now, both instruments showed a springtime loss of ozone of 30 per cent. By October 1987, the time of the Montreal Protocol, the lowest value measured was 125 Dobson units, indicating that more than half of the ozone had been destroyed, making the famous 'hole' in the ozone layer.

By then, the earlier data from Halley Bay had been published, and the NASA scientists responsible for Nimbus 7 had frantically gone over their data to find out why the satellite had failed to notice such a dramatic decline in ozone. It turned out that the instruments had indeed measured the decline of ozone over Antarctica – but the computers handling the data had been programmed to reject any measurements outside the 'normal' range of fluctuations! Echoing Lovelock's earlier experience, the data were rejected because they did not fit the model; as he wryly comments, rejecting anything outside your usual experience is hardly the way to make new discoveries. 'It was awful, absolutely awful. No way to do science.' There was one saving grace. Though the low readings were flagged as erroneous and ignored in the computer processing, the raw data had been saved, and when they were studied properly they not only confirmed the Halley Bay measurements, but showed that the ozone depletion covered the whole of the Antarctic continent. There really had been a drastic loss of ozone over the Antarctic region, although the stratosphere seemed to recover during the summer. This recovery was not the good news it might seem. It happened partly because ozone from lower latitudes, nearer the equator, moved in and filled up the hole – which meant that ozone levels were also falling, although less dramatically, outside the Antarctic circle.

The important lesson is never to have blind faith in your models, but to start with observations. Conan Doyle put it neatly in the Sherlock Holmes story 'A Scandal in Bohemia', where Holmes says:

> It is a capital mistake to theorise before one has data. Insensibly, one begins to twist facts to suit theories, instead of theories to suit facts.

Once the correct Nimbus 7 facts were combined with those of the Halley Bay team, an impressive scientific programme costing of tens of millions of dollars swung into action to find out what was going on. This included high-altitude flights by an ER-2 research aircraft (a converted U-2 spy plane), observations made by instruments carried on unmanned balloons, data from satellites, and many more observations from various ground stations. It soon became clear that the extreme cold of the winter stratosphere over Antarctica 'preconditioned' the ozone layer by providing a high-altitude haze of frozen particles on which chemical reactions involving chlorine can take place. Then, when the Sun returns in the spring, solar energy triggers the reactions that destroy ozone. It was confirmed that the chlorine came from the breakup of CFCs.[1] A similar process was later found to be at work in the Northern Hemisphere, over the Arctic, posing a much more immediate threat to densely populated regions of our planet. As with Rachel Carson's emotive claims about DDT, but with rather more scientific basis, once again the need for immediate action was stressed by the public fear of cancer – this time, skin cancer caused by the burning ultraviolet rays of the Sun. Perhaps it would be easier to get controls on greenhouse gas emissions if carbon dioxide caused cancer.[2]

History records that in the case of CFCs and ozone action was taken and was effective. This remains the best example of humankind identifying a threat to the planet – if you like, a threat to Gaia – and

1. For details of the chemistry, see John Gribbin, *The Hole in the Sky*, Corgi, 1988.
2. As is often the case, Lovelock has a different perspective from the alarmists. While he agrees that an increase in the amount of ultraviolet reaching the surface of the Earth is bad for 'light-skinned humans' and other specific organisms, including some plants and animals used for food, he doubts that it would have a marked detrimental effect on the biosphere at large. In his view, the threat of ozone depletion was not really a threat to Gaia, but 'only' a threat to human civilization.

taking effective action to counter the threat. Because CFCs are so long lived, even though they have essentially been banned it will take many decades before their effects cease to be felt, and even now there is still a hole in the ozone layer over Antarctica each spring. But the hole is getting smaller, and the threat of increased ultraviolet radiation affecting marine life around the Antarctic continent is getting less. As long ago as 1988, the broader importance of the concerted action to tackle the ozone problem was highlighted by John Maddox, then editor of the journal *Nature*, at a conference in London:

> The Montreal Protocol is a valuable precedent for the kind of convention that will in due course have to deal with greenhouse gases. Obviously the problem of global warming has much greater implications economically. Even so, and despite the uncertainties that persist, it is by no means too soon to begin embarking now on the negotiation and the kind of agreements that will be necessary to regulate the greenhouse problem.[1]

Slightly less optimistically, at the same meeting Robin Russell Jones, of St John's Hospital in London, concluded,

> I have no doubt that the issue of stratospheric ozone represents the litmus test of man's ability to prevent the ultimate degradation of Planet Earth. If this relatively simple problem cannot be solved, how can mankind hope to survive the vastly more complex problems that will arise in the centuries to come?

Well, the relatively simple problem *was* solved; whether we choose to solve the problem of global warming – there is little doubt that we *can* solve it – remains to be seen. One key difference is that in the case of ozone depletion, the United States, the most powerful economy in the world, took the lead; in the case of global warming the United States has so far been dragging behind.

Lovelock may not have been directly involved in the saga of the hole in the sky. But it did start to make him think more seriously about the threat to Gaia posed by human activities, even though he was not a 'doomwatch sort of person'. As early as 1973, at a meeting on 'The Ecology and Toxicology of Fluorocarbons' held in Andover,

1. See Jones and Wigley.

Maine, he had been the first scientist to point out (*before* the threat to the ozone layer was identified!) that there is another threat from CFCs – they are very effective greenhouse gases, and he estimated that by the time the concentration of CFCs in the atmosphere increased tenfold from the level of the early 1970s they would be making a significant contribution to global warming – there was, indeed, a conceivable way in which they could be a threat to the global environment.[1] Little attention was paid to the warning at the time – indeed, another scientist, Veerhabadrhan Ramanathan, is usually credited with blowing the whistle on CFCs as greenhouse gases, although he did not discover their effectiveness as infrared absorbers until 1975. The awareness of the threat posed by global warming was always in Lovelock's mind as he developed the concept of Gaia during the 1970s and 1980s. Even in his first Gaia book, published in 1979, he wrote, 'If, due to fossil fuel combustion, the level of carbon dioxide rose too rapidly for inorganic equilibrium forces to cope, the threat of overheating might become serious.'

Eight years later in 1987, at the time of the Montreal Protocol, the Gaia idea had been widely discussed, and taken up by some people as a kind of comfort blanket, the image of a benevolent, protective Earth Mother. This was far from what Lovelock had intended. Discussing the ozone problem, global warming and other environmental hazards with John Gribbin over coffee at a meeting in Berlin in November 1987, he said, 'People sometimes have the attitude that "Gaia will look after us." But that's totally wrong. If the concept means anything at all, Gaia will look after *herself*. And the best way for her to do that might well be to get rid of us.' That idea grew in his mind, as we shall see, to become the theme of his book *Revenge of Gaia*, published in 2006. But first, we should go back to the origins of the concept – the moment, which Lovelock can pinpoint precisely, when the idea of Gaia first occurred to him.

1. Lovelock also realized that there could be practical uses for CFCs as greenhouse gases, pointing out that if they were released in cold northerly airstreams moving down towards Florida, they would trap enough infrared energy to prevent frosts damaging the citrus crops. He got as far as initiating a patent application for the process, but that 'had to be dropped hastily' as the environmental implications of increased use of CFCs became clear.

7
The Revelation

The concept of Gaia came to Jim Lovelock 'suddenly, as a revelation' one afternoon in September 1965, when he was on a visit to the Jet Propulsion Laboratory in California. But the moment of epiphany, like so many 'sudden' insights, stemmed from a train of thought that had begun long before.

Early on in his relationship with JPL – which began in 1961 – Lovelock was invited to sit in on meetings of scientists and engineers planning experiments to search for life on Mars with a proposed unmanned lander called Voyager. He was surprised to find that all their ideas seemed to be based on finding the same kind of life that exists on Earth. Such experiments would be superb at finding traces of life in the Mojave Desert, near Los Angeles; but what if life on Mars were different? His comments ruffled the feathers of many of the experimenters, who felt that since life on Earth was the only kind they knew about, it was inevitably the only kind they could design experiments to search for. Pressed to say how he would design a general experiment to look for evidence of any kind of life, Lovelock, although at that time completely ignorant of the work of Alfred Lotka, replied that he would look for signs of entropy reduction. His off-the-cuff remark provoked a challenge to design an experiment that would do just that.

As Lovelock recalls, the challenge 'concentrated the mind wonderfully'. Starting from Erwin Schrödinger's famous book *What is Life?*, which set him on the right track, in a couple of days he came up with a list of possibilities. Top of the list was simply to analyse the composition of the Martian atmosphere to find out if it were close to the equilibrium state that corresponds to high entropy. High entropy corresponds to systems that have run down into equilibrium, where nothing interesting happens, and there is very little information; low entropy corresponds to systems which are being maintained out of equilibrium by a flow of energy, interesting things go on, and there is a lot of information. As we saw in Chapter 3, and as Lovelock realized, low entropy is a sign of life.

There were other possibilities on Lovelock's list, but this was his definitive life-detection experiment. It was simple, cheap and unambiguous. His idea was taken seriously at JPL, and as he was about to return to England he was urged to write a report on the idea for discussion on his next visit. He did so, and sent a slightly different version to the science journal *Nature*, under the title 'A Physical Basis for Life-detection Experiments'.[1] As it happened, this was the first scientific paper he had submitted for publication since leaving Mill Hill; it came straight back, with a note explaining that *Nature* did not consider papers sent from private addresses. When Lovelock contacted the editor and explained that he was the same person who had published a string of papers while working at Mill Hill, an exception was made and the paper duly appeared in print; but 'it was a sobering discovery that for the past twenty years my papers had got published not because of my brilliance as a scientist but simply because they came from a respectable address.'

On his next visit to JPL, in 1964, Lovelock met Dian Hitchcock, from NASA headquarters in Washington, who was intrigued by his idea and reported favourably on it to her superiors. He was invited to Washington to discuss it further, and found that his *Nature* paper had made a big impact among the administrators and scientists there. Aided by Hitchcock, who had a background in philosophy and an inside knowledge of how NASA worked, he drafted a proper proposal to put the idea into practice, and was soon 'astonished to find myself made acting chief scientist for the physical life-detection experiments on the planned *Voyager* mission'.

For an old science-fiction fan, this was a heady and exciting time. Lovelock returned to England with his mind buzzing with ideas and

1. *Nature*, vol. 207, pp. 568–70, 1965; the 'really definitive paper', says Lovelock, was published in *Advances in Astronautical Science*, vol. 25, pp. 179–93, 1969.

plans, but he was confronted with the serious problem of Helen's declining health. The local doctors had failed to diagnose her multiple sclerosis, and signs such as temporary episodes of dragging one foot while walking, or partial one-sided blindness, had been dismissed as nothing to worry about since they seemed to get better by themselves. Looking back, Lovelock feels that he was too ready to accept these comforting prognoses, and that he 'must have been a great trial to Helen', who never made a fuss about her condition, as he was so absorbed in his work. The great trial would continue for all the family, including Jim, for the next twenty-four years. 'It was just the way things were; we all had to get on with it.'

In the spring of 1965, in his new capacity as an acting chief scientist, Lovelock spent six weeks travelling around the United States, visiting contractors and working on proposals for the Voyager experiments. There was also a proposal to build an infrared telescope on White Mountain in California, specifically to look for the spectroscopic signatures of different gases, such as water vapour and carbon dioxide, in the atmospheres of other planets; but nothing ever came of it. Towards the end of this visit, on 14 July 1965 the Mariner 4 spacecraft on a Mars flyby sent back the first reasonably clear images of the Martian surface, revealing a seemingly barren desert with no signs of life; 'but,' says Lovelock, 'this only seemed to make the biologists more determined to find life there.' With Hitchcock, Lovelock prepared a paper on the planned Voyager experiment, which was submitted to the Royal Society on their behalf by Lord Rothschild. In a sign of how the scientific community would later react to the Gaia idea itself, the paper was rejected, much to Rothschild's disgust. But in a shining example of the way science ought to be done, Carl Sagan, whom Lovelock met at JPL, offered to publish it in his journal *Icarus*. 'Sagan disagreed with just about everything we said, but he felt strongly that the right thing to do was to have the idea discussed as widely as possible.'

In September 1965, Lovelock's career as a NASA chief scientist came to an end when Congress pulled the plug on funding for the Voyager mission.[1] It was essentially replaced by the cheaper Viking

1. The name was later resurrected for a completely different pair of probes which visited the outer planets of the Solar System.

probes, which would land on Mars in 1975. There was no room on Viking for Lovelock's experiments, and to his frustration the life-detection experiments planned for the mission were all of the kind designed to find life in the Mojave desert. It was some comfort that he was kept on the payroll to help develop instruments, but galling to think what might have been. Although asked to be a lead designer on the Viking instrument team, Lovelock had to decline, because it would have meant moving to California, which was impossible because of Helen's increasing disability.

As it turned out, though, there would have been no need for Lovelock's Voyager experiments anyway. That September, Lovelock was in an office at JPL one afternoon when one of the astronomers brought in some data that had been gathered by a team at the Pic du Midi Observatory in France. They had used infrared instruments – exactly as Lovelock had suggested – to identify the composition of the atmospheres of our nearest neighbouring planets. They found that both Mars and Venus have atmospheres dominated by carbon dioxide, with only traces of other gases. They are in equilibrium, with high entropy. To Lovelock, the implication was clear – there is no life on Mars (or, for that matter, on Venus), and the life-detection experiments planned for the Viking probes were a waste of effort. Or were they? The test of a good scientific theory is whether it makes predictions that are borne out by experiment; Gaia theory, as it became, predicted that the Viking probes would find no trace of life on Mars; the experiments confirmed that prediction.

It was the discovery that Earth is the odd one out of three planets, the only one with an atmosphere rich in oxygen, that made something click in Lovelock's mind and led to the concept of Gaia. Oxygen must always be being destroyed in the air by reactions with other substances, such as methane, or simply by burning. What sustains its presence? What happens to the carbon dioxide and other products of combustion? The gases come from living things; they are absorbed by living things. 'Suddenly, like a flash of enlightenment, it came to me. *Living things must be regulating the composition of the atmosphere.*' Lovelock revealed his insight to Sagan and Hitchcock, blurting it out to his companions as it came to him; but it made little impression at the time. Then, Sagan told him about the 'faint young Sun paradox', which says that although according to our understanding of astrophysics the Sun was about 25 per cent cooler when it was young, the temperature of the Earth was never so low that the oceans froze solid. This reminded Lovelock of something familiar from his medical background – the way a warm-blooded animal maintains its body at a constant temperature for a wide range of outside temperatures, an example of homeostasis. 'The image of the Earth as a living organism able to regulate its temperature and chemistry at a comfortable steady state emerged in my mind.'

The first scientific paper setting out these ideas was written by Lovelock in collaboration with a colleague at JPL, C. E. Giffin, in 1968, from the perspective of the search for life on other planets. It starts from the understanding that 'it is a property of life to reduce its internal entropy,' but offers a new view of the maximum size of a 'unit of life' and what is meant by 'internal'. Since an animal exports disorder into the atmosphere of the Earth, 'it might seem pointless therefore to seek evidence of life by looking for order in the chemical composition of the atmosphere.' But:

> If instead of individual living organisms, however, the planetary ecosystem itself is regarded as the maximum unit of life, the problem resolves. In an ecosystem, the atmosphere can have an ordered role as the conveyor belt for products between, for example, the plant and animal kingdoms of Earth, or their analogues elsewhere. With this large unit, *the atmosphere is an internal component of the living system and the environment is now space, to which disorder is rejected in the form of degraded solar energy*.[1]

The paper also points out that the dominant component of the Earth's blanket of air, nitrogen, although 'often regarded as if it were a stable permanent product of primitive outgasing', left over from when the Earth was young, actually has a lifetime in the atmosphere of less than a hundred million years, and must be being constantly renewed, probably by biological processes. 'Whatever its past may have been, the air now is revealed as a complex biological contrivance.' And in an almost throwaway aside, 'it may not therefore be an unreasonable speculation to consider the possibility that the Earth's

1. Our emphasis.

climate is also maintained at or near an optimum for the ecosystem' by biological control.

The bones of Gaia theory were in place by 1968. But it took a long time to flesh out the image. Over the next few years, Lovelock was busy with his consultancy work, and, with his three eldest children having left home, preoccupied at home with helping Helen to cope. The fact that 'home' was also his main place of work at least allowed him to fulfil the dual role, but he was left with little time to develop his image of the Earth as a living organism. He did present the basic idea of a self-regulating Earth System at a few scientific meetings, including one at Princeton University in 1968, where he met the biologist Lynn Margulis, an expert on the structure of the cell and the way different components of the cell have come together to make a single entity. She later became a close collaborator in developing Gaia theory, but they had no discussions at that first meeting, where, as the most junior scientist present, she was concentrating on preparing a report of the proceedings for publication. Looking back on that meeting, Lovelock gives an interesting insight into his own self-image. 'I wasn't surprised that my ideas didn't attract attention. Young scientists like us [himself and Margulis, who had been born in 1938] were expected to be seen and not heard.' The 'young scientist' was already 49 in 1968; but even today he is diffident and uncomfortable about pushing himself, or his ideas, forward. That was one reason why the collaboration with Margulis worked so well – a more forceful person with a more American approach to promoting Gaian ideas, and working within the conventional academic system at Boston University, she was the ideal foil.

Margulis puts this diffidence in perspective. When they first met properly, she hadn't even known that Jim had been at the Princeton meeting, and he had to remind her by showing her the Proceedings volume. Then she recalled that Jim had been the anonymous person who asked a question at one of the sessions and the speaker 'had shut him up so vehemently that [Lovelock] remained silent for the rest of the meeting'.

Their collaboration began, initially by mail, in the summer of 1970. Margulis had independently become intrigued by the oddity of the Earth's oxygen-rich atmosphere, and asked her former husband, Carl

Sagan, who she ought to discuss the puzzle with. Sagan knew just the man, and put her in touch with Lovelock. They first met properly in 1971, at Boston University; over the next few years Boston became the first stop on Lovelock's regular visits to the United States. One of the areas they investigated together was the way oxygen reacts in the mitochondria inside cells – the powerhouses of the cell. The fossil evidence shows that this process has been going on over geological timescales, but if the oxygen concentration fell below 10 per cent the reaction would not work and cells would die. On the other hand, Lovelock had turned his attention to the puzzle (which he originally noticed in 1968), that the level of oxygen in the air could not go much above 20 per cent, or vegetable matter would burn. Some aspect of the living system, they realized, kicks in to push the level up or down as needed to maintain conditions suitable for life.

This was the beginning of a fruitful collaboration in which Margulis found Lovelock to be 'profoundly creative, intellectually mischievous, and kind'. On one of his visits, a colleague asked if he could record a discussion with Lovelock, which ended up lasting from 10 a.m. until 4 p.m. – at which point they discovered the recorder wasn't working. Although Lovelock was leaving town that evening, he 'just calmly sat down and re-recorded it'. The tape provided the basis for one of the first Gaia papers, published in *Icarus* in 1974.

Lovelock's travels in the early 1970s also took in the US National Center for Atmospheric Research (NCAR) in Boulder, Colorado, where he used to stop over on visits to JPL, and where he discussed his ideas with atmospheric scientists and climatologists. By then, Gaia had been named – after the Greek goddess of the Earth – by the novelist William Golding, who was a neighbour and friend of the Lovelocks in Bowerchalke, saying that a big idea deserved a big name.[1]

The concept may have been developing slowly in Lovelock's mind, with little attention paid by the scientific community or the world at large, but it was clearly an idea whose time had come. It is very often the case in science that different people come up with the same big idea (or parts of the same big idea) at more or less the same time.

1. As Golding realized, the Greek word *gaia* is also the root for words such as geometry and geology.

The two biggest examples are the way Robert Hooke came close to developing 'Newton's' theory of gravity before Newton did, and the way Alfred Wallace came up with the theory of evolution by natural selection independently of Darwin. By the early 1970s, several versions of the idea of the Earth as a single organism were being aired, although none of them took off.

The key concept is homeostasis, a term first used biologically in the 1930s[1] to describe the way the composition of the blood of a mammal stays the same in spite of changes in the surrounding environment, but which biologists now use more generally to refer to the way an organism maintains a more or less constant internal state. The fact that, for example, our bodies maintain themselves at roughly the same temperature all the time is so obvious – literally, so natural – that it is only when you stop and think about it that you realize how truly remarkable it is compared with the behaviour of a non-living system. Homeostasis applies to everything from individual cells to whole bodies – and, according to Gaia theory, the whole planet. In his superb book *An Introduction to the Study of Man*, published in 1971, J. Z. Young,[2] then Professor of Anatomy at University College London, wrote:

> The key to understanding the significance of the activities of plants, mammals, and men is the recognition of their homeostatic nature, and the fact that they tend to preserve the continuity of life. This is obvious enough for many actions, for example, breathing and feeding. The difficulty is to recognize the significance for homeostasis of other activities such as the complicated operations of mating or the apparently self-destructive activities of growing old. In order to do this it is necessary to focus attention on the fact that the stability of the organization is maintained not merely for short periods but for hundreds, thousands, and millions of years. *The entity that is maintained intact, and of which we all form a part, is not the life of any one of us, but in the end the whole of life upon the planet.*[3]

1. W. B. Cannon, *The Wisdom of the Body*, Norton, New York, 1932.
2. Usually known by his initials, which stand for John Zachary, rather than his full name. He was a descendant of Thomas Young, famous for developing the wave theory of light.
3. Young's emphasis.

This comes close to recognizing Gaia, although Young stops short of pointing out that life regulates the composition of the atmosphere and other physical features of our planet.

Around the same time, another insight came from the physician Lewis Thomas, originally presented in an article in the *New England Journal of Medicine* and later (in 1974) in a collection of essays under the title *The Lives of a Cell*. Like Margulis, Thomas was intrigued by the way the different components of a living cell work together to maintain homeostasis. Margulis pioneered and promoted the idea, now part of conventional biological wisdom, that many of these components (sometimes known as organelles) are the descendants of what were, long ago, independent organisms that 'learned' (through evolution by natural selection) to cooperate with one another symbiotically to make a stable environment within the membrane that makes up the wall of a cell. What were once free-living bacteria are now incorporated within animal and plant cells, where they carry out specific tasks, such as converting sunlight into useful energy. The cell as a whole benefits from the division of labour; the individual components benefit from being in a stable environment.[1] Or as the zoologist Thomas Cavalier-Smith, of Oxford University, has put it, in cells 'there is a mutualistic cooperation of genomes, membranes, skeletons and catalysts that together makes a physically and functionally coherent unit.'[2] In Lovelock's view, just such a 'mutualistic cooperation', but on a larger scale, is what makes Gaia alive.

It is no coincidence that Thomas should have applied his biological background to thinking about the Earth System at about the time Lovelock did. Just as Lovelock's revelation about the nature of Gaia came from his work with NASA and thinking about the Earth as a planet, so Thomas's insight was a direct product of the space age – in particular, the pictures of the Earth taken by Apollo astronauts as they orbited the Moon:

1. And at one step further up the ladder of complexity, a body like ours is made up of many cells, which are specialized to carry out different tasks (as muscle cells, or liver cells, or whatever), with the whole community of cells benefiting as a result.
2. *Nature*, 15 March 2007, p. 257.

Viewed from the distance of the moon, the astonishing thing about the earth, catching the breath, is that it is alive. The photographs show the dry, pounded surface of the moon in the foreground, dead as an old bone. Aloft, floating free beneath the moist, gleaming membrane of bright blue sky, is the rising earth, the only exuberant thing in this part of the cosmos. If you could look long enough, you would see the swirling of the great drifts of white cloud, covering and uncovering the half-hidden masses of land. If you had been looking for a very long, geologic time, you could have seen the continents themselves in motion, drifting apart on their crustal plates, held afloat by the fire beneath. *It has the organized, self-contained look of a live creature, full of information, marvelously skilled in handling the sun.*[1]

Writing for a non-scientific audience, Thomas avoids the word 'entropy', but it is no accident that he refers to the living Earth as 'full of information'. The word 'membrane' is not used lightly, either.

Membranes are what enable biological systems to produce pockets of order in a sea of disorder. A living organism has to have somewhere to store energy (or food) and hold on to it, releasing what it needs when and where it needs it. A cell does this within the safety of its surrounding membrane; so do the organelles inside a cell, which have their own membranes – part of their heritage as descendants of free-living bacteria. All organisms eat, and all excrete waste. The Earth's 'food' is high energy radiation from the Sun, which is relatively low in entropy;[2] it's waste is low energy infrared radiation, relatively high in entropy. Life feeds off the flow of energy from the Sun; but it can only do so because it has somewhere to trap part of that energy flow. 'When the earth came alive,' says Thomas, 'it began constructing its own membrane,' by modifying the atmosphere around the planet and constructing, among other things, the ozone layer. More than thirty years ago, long before the discovery of the hole in the sky or the late twentieth century debates about the greenhouse effect, he said that:

1. Our emphasis.

2. In entropy terms, what matters is the temperature of the radiation, above the absolute zero of temperature, –273°C. In round terms, the surface of the Sun radiates energy at about 6,000 degrees above absolute zero, while the Earth radiates energy at a temperature of 300 degrees above absolute zero.

we are safe, well ventilated, and incubated, provided we can avoid technologies that might fiddle with that ozone, or shift the levels of carbon dioxide . . . taken all in all, the sky is a miraculous achievement. It works, and for what it is designed [we would prefer, 'has evolved'] to accomplish it is as infallible as anything in nature.

Thomas makes it clear that he is not simply using the cell as an analogy:

I have been trying to think of the earth as a kind of organism, but it is no go. I cannot think of it this way. It is too big, too complex, with too many working parts lacking visible connections. The other night, driving through a hilly, wooded part of southern New England, I wondered about this. If not like an organism, what is it like, what is it *most* like? Then, satisfactorily for that moment, it came to me: it is *most* like a single cell.

Developing the theme in terms of the relationship between the different kinds of life on Earth and the different organelles in a cell, he says,

We do not have solitary beings. Every creature is, in some sense, connected to and dependent on the rest . . . We are part of the system. One way to put it is that the earth is a loosely formed, spherical organism, with all its working parts linked in symbiosis.

And those working parts include the air that we breathe, the ozone layer and the other physical systems of our planet.

We have gone into Thomas's insight in some detail because the image of the Earth as a single living cell is, we believe, the best and clearest to keep in mind as we continue the story of the development of the Gaia concept.

Gaian ideas were in the air in the early 1970s, and it was inevitable that sooner or later someone would develop them into a full-blown scientific theory. The fact that it happened sooner rather than later, and that the someone was Jim Lovelock, owed much to his unusual background as an interdisciplinary scientist with an understanding of engineering principles – Thomas could never have taken the key step because he wasn't a hands-on inventor – and the collaboration with Margulis, an expert on the symbiotic systems of the cell, provided the last piece of the puzzle. Even so, with both of them having little time to spare from their real jobs to devote to Gaia, it took a long time for the concept to develop, and the debate about Gaia only really took off when it entered the public arena.

In 1971, Lovelock attended a meeting in New Hampshire on atmospheric chemistry, where he spoke about his measurements of CFCs in the atmosphere – the talk which alerted Lester Machta, and through him Sherry Rowland, to the potential threat to the ozone layer. To entertain the conference participants after dinner one evening, he was also asked to give a fifteen-minute talk on Gaia – the first time the idea had been aired for atmospheric scientists. The talk was published a year later in the journal *Atmospheric Environment*. Lovelock and Margulis followed this up with a couple of more technical papers, one (with Lovelock as lead author) on homeostasis and the biosphere,[1] the other (with Margulis as lead author) on biological modulation of the atmosphere. The idea still failed to attract attention, but in 1974, to his astonishment, Lovelock's lifetime work as a practical scientist and inventor was recognized by his election as a Fellow of the Royal Society, the ultimate seal of approval from the British scientific establishment. No longer could he pretend, even to himself, that he was just 'a young scientist who should be seen and not heard'.

But the honour made little difference when it came to promoting Gaia. Lovelock became convinced by Margulis that microorganisms are at the heart of Gaia, but neither of them seemed able to convince anyone else that there was anything in the idea of Gaia. 'They could not prove us wrong but they were sure in their hearts that we were.' But nor did the publications rouse much opposition. 'They just ignored us in the hope that we might go away.' Things began to change in 1975, when Lovelock, in collaboration with Sidney Epton, a chemist working for Shell, published an article titled 'The Quest for Gaia' in the popular science magazine *New Scientist*.[2]

In that article, Lovelock and Epton point out that on the conventional view that 'life exists only because material conditions on Earth

1. This article, published in the journal *Tellus* in 1973, has been described by climatologist Stephen Schneider as 'the classical scientific article on Gaia'.
2. It appeared in print on 6 February 1975.

happen to be just right for its existence', the implication is that 'life has stood poised like a needle on its point' for more than 3,500 million years. 'If the temperature or humidity or salinity or acidity or any one of a number of other variables had strayed outside a narrow range of values for any length of time, life would have been annihilated.' The alternative is that 'life defines the material conditions needed for its survival and makes sure that they stay there.'

They use the example of the faint young Sun paradox, suggesting that long ago the Earth had an atmosphere rich in greenhouse gases that kept it warm when the Sun was cool, and that the actions of life reduced the concentration of these gases over the aeons, allowing the Earth to maintain a stable temperature while the Sun warmed. They discuss the Martian life-detection experiments and chemical thermodynamics, and conclude 'the atmosphere [looks] like a contrivance put together cooperatively by the totality of living systems.' But their most dramatic statements are reserved for the role of humankind:

> If one showed a control engineer the graph of the Earth's mean temperature against time over the past million years,[1] he would no doubt remark that it represented the behaviour of a system in which serious instabilities could develop but that had never gone out of control. One of the laws of system control is that if a system is to maintain stability it must possess an adequate variety of response, that is, have at least as many ways of countering outside disturbances as there are outside disturbances to act on it. What is to be feared is that Man-the-farmer and Man-the-engineer are reducing the total variety of response open to Gaia . . .
>
> We are sure that Man needs Gaia, but could Gaia do without Man?

The rhetorical question set the seal on what was obviously, whatever its scientific merits, a marketable idea. Lovelock promptly received twenty-one invitations from publishers to write a book about what the *New Scientist* article called 'the Gaia hypothesis'; he chose Oxford University Press because he liked the editor they sent to talk to him (Peter Janson Smith) and because of their impeccably respectable credentials as an academic press. The book, *Gaia: A New Look at Life on Earth*, was mostly written during the Lovelocks' summer visits

1. This graph oscillates several times between Ice Age and Interglacial conditions.

to Adrigole. It was finished at the beginning of 1977, and published in 1979.

This was just about the last use the Lovelocks were able to make of the Adrigole cottage; they had to give up these excursions in 1978 because Helen's increasing disability made it impossible for her to travel to the west of Ireland.[1] There were other domestic changes in the second half of the 1970s. It was after the Lovelocks returned from Ireland to Bowerchalke in the late summer of 1976 that an incident occurred which made them realize how much the village had changed. They received a complaint that their garden was rather scruffy, and would detract from Bowerchalke's chances in the Best Kept Village competition. Worse, the censure came from a man who had only recently retired to the village, and had no roots there. The incident opened their eyes to the way the village had changed during the 1970s. The village cricket team, the school and the pub had all gone, and they were now among the last representatives of the old way of life, surrounded by 'a gentrified nest of middle-class strangers'. It was far from being the village it had been in the 1960s, let alone the village of Lovelock's boyhood dreams, and with the children grown up it was time to move on. Helen's disability was a contributing factor. The uphill path from the road to the house, and the stairs inside, were increasingly difficult for her.

Characteristically, they moved further away from the centres of urban civilization, and in April 1977, with Jim's book safely in the hands of its publishers, purchased Coombe Mill, on the border of Devon and Cornwall. Laid out mostly on one floor with a level approach from outside and surrounded by its own land (which they added to over the years), it was an ideal retreat. The house in Bowerchalke, soon to be known as Lovelocks, sold for almost the same price they paid for Coombe Mill and its land. The property was remote enough to discourage casual visitors, and a sign Jim placed on the gate of the property, reading 'Coombe Mill Experimental Station', did wonders to deter strangers, who 'were never quite sure what kind of experiments went on'. This completed what might well have turned

1. But the cottage is still in the family – Jim and Helen sold it to their son-in-law, Michael Flynn.

out to be, with Lovelock approaching his sixtieth birthday, a quiet semi-retirement.

The years from 1977 to 1979 were indeed the quietest of all Lovelock's life as an independent scientist. Although he still carried out consultancy work for Shell, Hewlett Packard, the Ministry of Defence and NOAA, travel was much more difficult, with even London some five hours away by car and train. Coombe Mill needed a lot of attention, with extra rooms being built for Jim's laboratory and Helen's office, and fourteen acres of newly purchased adjoining land to look after. But in the year Lovelock turned 60, his first book was published. Things would never be the same again.

In the introduction to the first edition of *Gaia*, Lovelock said that it was written 'primarily to stimulate and entertain'. It certainly did both, reaching a wide popular audience and, perhaps partly because of its popular success, provoking a sometimes furious response from scientists. 'The biologists were the worst,' says Lovelock. 'They spoke against Gaia with the kind of dogmatic certainty I hadn't heard since Sunday School. At least the geologists offered criticisms based on their interpretation of the facts.'

Part of the problem was that, having more or less given up on the scientific community taking any interest in Gaia, Lovelock had written his book for a lay audience, imagining that he was composing 'a long letter about Gaia to a lively, intelligent woman' who had no background in science.[1] The tone was set from the opening paragraph of the introduction:

> This book . . . is about a search for life. And the quest for Gaia is an attempt to find the largest living creature on Earth. Our journey may reveal no more than the almost infinite variety of living forms which have proliferated over the Earth's surface under the transparent case of the air and which constitute the biosphere. But if Gaia does exist, then we may find ourselves and all other living things to be parts and partners of a vast being who in her entirety has the power to maintain our planet as a fit and comfortable habitat for life.

1. This is not quite the sexist remark it may seem today; Lovelock had in mind as a role model George Bernard Shaw's *An Intelligent Woman's Guide to Socialism, Sovietism and Capitalism*.

Perhaps the biggest mistake Lovelock made in presenting his case comes towards the end of that introduction:

> Gaia [is] a complex entity involving the Earth's biosphere, atmosphere, oceans, and soil; the totality constituting a feedback or cybernetic system which seeks an optimal physical and chemical environment for life on this planet. The maintenance of relatively constant conditions by active control may be conveniently described by the term 'homeostasis'.

If only he had left out the bit about 'optimal' conditions, he would have saved himself a lot of bother, as he now recognizes. 'I should have said, "a suitable environment for life".'

But there are three key points in that paragraph which we can use to illustrate what this first Gaia book was all about – feedback, the physical environment, and the chemical environment.

Feedback occurs when some activity produces an action which literally 'feeds back' on the original activity to make it either more effective (positive feedback) or less effective (negative feedback). A familiar example of positive feedback is the howl that comes from a loudspeaker when a microphone that is connected to an amplifier that is itself connected to the speaker is placed close to the speaker. Any tiny noise emerging from the speaker is picked up by the mike, amplified by the amplifier, and fed back, louder, to the mike. A very few loops around the system produces an ear-splitting screech. Negative feedback is used in speed-control systems in cars. If the speed is set to, say, 50 mph and the car encounters an upward slope, the car will start to slow down, so the computer will increase the power going to the wheels to compensate and cancel out the slowing down; as the road levels off again, the speed will increase, and the computer will reduce the power to the wheels to bring the speed back down to 50 mph.

We can imagine both positive and negative feedback occurring in imaginary climate systems, mental models of the world. Such models can be useful in helping us to get an idea of how physical systems work, although they are not as useful as computer models in which the various interactions between components, including feedbacks, are calculated accurately in line with the laws of physics. Imagine, for example, a planet in the same orbit as the Earth on which there are oceans, continents, ice caps and nothing else. If for any reason the temperature goes up (perhaps the Sun gets brighter) and some of the ice melts, it will expose dark land in the place of shiny white ice, changing the reflectivity of the Earth (its albedo). The dark land will absorb more heat from the Sun than the ice did, which will make the temperature go up more, melting more ice in a positive feedback. Then again, if the temperature goes up, more water may evaporate from the oceans, making more clouds. The white clouds will reflect away more heat from the Sun, cooling the planet below and restoring the status quo in a negative feedback. We are not saying that either of these effects will dominate on a real planet like the Earth; that is a complex problem which can barely be tackled by the best computer models available today. All we are saying is that these are imaginary models of how positive and negative feedback could work on a real planet. Lovelock's contention is that although both kinds of feedback are at work on the Earth, for some four billion years Gaian mechanisms of negative feedback have dominated to ensure 'relatively constant conditions'.

The first example of this Gaian mechanism does indeed involve temperature, and the faint young Sun paradox. When Carl Sagan and his colleague George Mullen, at Cornell University, drew attention to this puzzle in the early 1970s they suggested that the early Earth remained unfrozen even when the Sun was cool because of a 'super greenhouse effect' caused by the presence of large amounts of methane and ammonia in the atmosphere of the young planet. They chose those gases partly because they are similar to the hydrogen-rich compounds seen in the atmospheres of the giant planets in the outer part of our Solar System, and partly because they are good greenhouse gases. But today it seems much more likely that the atmosphere of the early Earth was rich in carbon dioxide – a gas produced copiously in volcanic activity – like the atmospheres of our near neighbours Venus and Mars. In those terms, the puzzle isn't why the early Earth was warm, but why the present-day Earth isn't superhot, like Venus. How was the amount of carbon dioxide in the air reduced steadily as the Sun warmed, to keep the temperature at the surface of the Earth in the range suitable for life?

It is relatively easy to come up with an imaginary model of how this might happen today, through the action of photosynthesis. Photosynthesizing organisms – in global terms, the most important ones are the tiny but numerous phytoplankton that live in the oceans – feed off carbon dioxide, and by and large they thrive when there is more carbon dioxide around. Even though the gas is constantly being released by volcanoes, they take it out of circulation by converting it into carbonates in their shells, which fall to the sea bed and get buried when the organisms die. So if the carbon dioxide concentration of the atmosphere goes up a little, then over thousands and millions of years photosynthesizing organisms will be more active and take more of it out of the air, preventing any dramatic rise in temperature. If the carbon dioxide concentration falls a little, then photosynthesizers will be less prolific, and the carbon dioxide concentration will build up, preventing any dramatic cooling. Negative feedback at work.

But none of this would be of any 'use' (from a Gaian perspective) if the biological material simply died and decayed where it fell, absorbing oxygen from the air and re-forming carbon dioxide. Free oxygen only exists in the air in substantial quantities today because carbon has been buried in the rocks;[1] it is the interaction between biological and geological processes that allows free oxygen to exist on Earth. Part of the problem with the present rapid buildup of carbon dioxide in the air caused by human activities is that we cannot wait for thousands, let alone millions, of years for such processes to come to our aid.

But the kinds of phytoplankta involved in these processes only evolved relatively recently during the history of life on Earth, so they cannot have been responsible for drawing carbon dioxide out of the air when the Earth was young. What could? Lovelock's answer is the blue-green algae, single-celled organisms which, fossil remains show, were among the first living things on Earth. In a few places today, most notably at Shark Bay in Western Australia, there are mounds of mushroom-shaped rock, a metre or so across, produced in shallow water by colonies of blue-green algae that have built up these structures, rich in carbonates and silicates, in which they live. These structures are known as stromatolites; although they are rare today,

1. Both in the obvious form of coal and oil and in the slightly less obvious form of limestone and chalk.

their fossil remains are found widely in rocks more than three thousand million years old. There is no doubt that the biological activity involved in making these ancient stromatolites took carbon dioxide out of the air and locked it up in rocks. The question is whether they were involved in the kind of Gaian feedbacks that Lovelock invoked in his first book.

In that book, however, he was concerned not so much with proving that Gaia was real, as with presenting a plausible case that Gaia *might* be real – the difference between a theory and a hypothesis. With that in mind, and realizing that no single temperature control process would suffice to keep Gaia healthy (we have already come across one other such process, the albedo effect), he drew on his medical training to make an analogy with the way the human body controls its own temperature, and introduced the idea of 'geophysiology'. Our temperature control mechanism involves at least five feedback systems, and there is no reason to think that Gaia would be any less complex.

The processes which control the temperature of your body include two distinct kinds of shivering – one associated with the skin and the other with the core of your body – sweating, generating heat by 'burning' food, and adjusting the diameter of blood vessels to increase or decrease the flow of blood near the skin. As Lovelock put it, 'our ability to sweat or shiver, to burn food or fat, and to control the rate of blood flow to our skin and limbs, is all part of a cooperative system for the regulation of our core temperature over an environmental range from freezing to 105°F (40.5°C).' But, 'even though we may find evidence of a Gaian system of temperature regulation, the disentangling of its constituent loops is unlikely to be easy.'

Lovelock then goes on to discuss the way Gaian mechanisms could control the chemical composition of the environment. In the early 1970s, he had become interested in the sulphur cycle – the natural processes which move sulphur around in the environment. Sulphur is an essential element in living organisms, found in some amino acids and many proteins, so these processes must involve life. A few years earlier, scientists had discovered that the amount of sulphur, in various compounds, being washed off the land and into the sea each year was more than could be accounted for by all the sources of sulphur on land. It seemed obvious that sulphur was somehow being recycled from the sea back onto the land, in quantities of hundreds of millions of tonnes a year, without which life on the continents would suffer. The chemists guessed that it was returning in the form of hydrogen sulphide, the gas which gives the familiar 'bad eggs' smell to vapours escaping from stagnant ponds, where organic remains are decomposing. Lovelock didn't go along with this. For one thing, the seashore doesn't smell of bad eggs; it smells fresh and invigorating.[1] That smell, he knew, is associated with another sulphur compound, dimethyl sulphide (DMS). For another thing, hydrogen sulphide reacts strongly with oxygen dissolved in seawater, so any bubbles of the gas formed beneath the sea would be converted into non-volatile compounds and never reach the surface. The clincher is that many organisms have evolved a mechanism for adding methyl groups (essentially, a molecule of methane with one hydrogen atom removed) to unwanted substances to get rid of them. Elements such as sulphur, mercury, antimony and arsenic are difficult to get rid of in their pure state, but their methyl compounds are much more volatile and can be released as gases when they are not wanted. Seaweeds and many other species of marine algae release DMS in large quantities.

All these ideas crystallized in Lovelock's mind in the summer of 1971, when the sulphur cycle was among the topics discussed at the conference in New Hampshire where he first presented his measurements of the concentration of CFCs in the air – and, as it happened, some measurements of sulphur compounds. It seemed much more likely to him that the undiscovered component in the sulphur cycle was DMS produced by organisms in the sea and returning to land in air currents. When he submitted his proposals for the *Shackleton* voyage later that year, he included both DMS and CFCs as the targets for his observations. Both were rejected, as we have seen, and Lovelock undertook the voyage with a free passage but paying for his own experiments and his own return air fare from Montevideo.

Although the observations made on the *Shackleton* voyage did show that DMS is present in samples of seawater even from the middle of the ocean, it turned out that this is not the most important source of the gas. Building on Lovelock's pioneering work, Peter Liss,

1. The smell often wrongly attributed to 'ozone'.

of the University of East Anglia, showed that the main sources of DMS lie in the cool seas at high latitudes, where 'one finds certain kinds of algal seaweed with an astonishingly efficient mechanism for extracting sulphur from sulphate ions in the sea and converting it to dimethyl sulphide.'[1] DMS produced in the shallow waters of continental shelves is, in fact, the main source of the sulphur returning to the continents to keep the sulphur cycle in balance. 'The paper Liss wrote on DMS was one of the key papers in Gaian research,' says Lovelock.

But why should the marine organisms 'care' about recycling sulphur to the land? Using the metaphor of selfishness promoted by Richard Dawkins in his book *The Selfish Gene*, what's in it for them? After all, it takes energy to extract sulphate from the sea and turn it into DMS which can blow away in the wind, and in purely selfish terms organisms would seem to be better off using that energy to grow and reproduce. Although, as we shall see in the next chapter, there is more to the DMS story than meets the eye, all Lovelock could say in his first book is that:

> The biological methylation of sulphur appears to be Gaia's way of ensuring a proper balance between the sulphur in the sea and on the land. Without this process, much of the soluble sulphur on land surfaces would have been washed off into the sea long ago and never replaced, thus disturbing the delicate equations between the environmental constituents needed for the maintenance of living organisms.

Such a disturbance would be bad news for life on land, perhaps; but what difference would it make for life in the sea? Why should life in the sea seemingly act altruistically for the benefit of life on land? It is easy to see why statements like this infuriated evolutionary biologists such as Dawkins, who responded vigorously to the idea that seemed to them to be an essential component of Gaian thinking – that one species could act for the benefit of other species and get nothing in return.

It was Lovelock's contention that the whole is greater than the sum of its parts, and in the first edition of *Gaia* he quoted his colleague Lynn Margulis as saying:

1. Lovelock, *Gaia*.

Each species to a greater or lesser degree modifies its environment to optimize its reproduction rate. Gaia follows from this by being the sum total of all these individual modifications and by the fact that all species are connected, for the production of gases, food and waste removal, however circuitously, to all others.

It's one of life's little ironies that Dawkins' most scathing attack on Gaia came in his book *The Extended Phenotype*, published in 1982, which promotes his idea that genes, the basic unit of natural selection, can indeed have an influence beyond the confines of the organism which they are part of. But he is thinking on a much smaller scale than Lovelock and Margulis. 'The farthest action at a distance I can think of,' he says, 'is a matter of several miles, the distance separating the extreme margins of a beaver lake from the genes for whose survival it is an adaptation,' because:

A gene in a beaver which . . . causes a larger lake to come into existence, can directly benefit itself by means of its lake. [Genes] causing smaller lakes are less likely to survive.

But although this is real science, Lovelock's book is dismissed by Dawkins as part of 'the pop-ecology literature'. While Lewis Thomas's likening of the world to a living cell 'can be accepted as a throwaway poetic line', Lovelock 'really means it' and 'cannot be ignored'.

Dawkins' main objection to the Gaia hypothesis as originally formulated is that it does not take proper account of evolution by natural selection. In order for Gaia to have evolved, said Dawkins:

There would have to have been a set of rival Gaias, presumably on different planets. Biospheres which did not develop efficient homeostatic regulation of their planetary atmospheres tended to go extinct. The Universe would have to be full of dead planets whose homeostatic regulation systems had failed, with, dotted around, a handful of successful, well-regulated planets of which Earth is one . . . In addition we would have to postulate some kind of reproduction, whereby successful planets spawned copies of their life forms on new planets.

Dismissing these notions as so ludicrous that even Lovelock could not possibly believe them, Dawkins concludes that, 'Obviously he simply did not see his hypothesis as entailing the hidden assumptions that I think it entails.' Lovelock 'might maintain that Gaia could evolve her global adaptations by the ordinary processes of Darwinian selection acting within the one planet. I very much doubt that a model of such a selection process could be made to work.'

Writing in the *Co-Evolution Quarterly*, the American biologist Ford Doolittle, now of Dalhousie University in Canada, expressed similar sentiments, pointing the finger at the kind of claims implicit in Margulis's description of Gaia:

It is not the difficulty of unravelling Gaian feedback loops that makes me doubt her existence. It is the impossibility of imagining *any* evolutionary mechanisms by which these loops could have arisen or now be maintained.

Although stung by such criticisms – not so much by the fact of being criticized, as by the hurtful tone adopted by some of his critics – Lovelock was still sure he was right. If it was impossible for Doolittle to imagine any evolutionary mechanisms by which feedback loops could maintain the homeostatic regulation systems of Gaia, and Dawkins doubted that a model of such a selection process could be made to work, he would prove them wrong.

Part of the problem of making Gaia credible was that she was embraced wholeheartedly by 'New Agers' and hippy types who misinterpreted what Lovelock was saying. With friends like that, Gaia was assumed by many biologists to be nothing more than a semi-religious view of Mother Earth. Years later, Lovelock discovered that even so noted an evolutionary biologist as John Maynard Smith had dismissed the whole idea of Gaia without even reading Lovelock's book, let alone any of his scientific papers. By then, thanks to the work stimulated by the stinging criticisms of the first popular presentation of Gaia, the hypothesis had been refined into a theory, and Maynard Smith was among the critics who changed their tune. Looking back, Lovelock now says that he is deeply grateful to those critics who did highlight flaws in his argument, and pointed him in the right direction for the next phase of his work. 'That's the way science is supposed to work, and the process worked very well; but I do think some of them might have phrased their criticisms less harshly. Mind you, that word "impossible" was like a red rag to a bull. Right from my earliest days in science, I never took it at face value when some senior bloke said something was impossible.'

In fact, by the time *The Extended Phenotype* appeared Lovelock was already developing a model which would answer the key criticisms of Doolittle and Dawkins; but it would not be until the end of the 1980s that he really began to win converts among the evolutionary biologists. This was partly because of the difficulty of finding ways to make Gaia scientific, but also because of serious health problems affecting Lovelock himself, and the inevitable progress of Helen's illness.

8

What Doesn't Kill You Makes You Strong

In science, theories are tested by making predictions which can be compared with the outcome of experiments and observations. If the theory fails the tests, it is dead; if it passes, it emerges stronger than before. Some people succumb to illness without putting up much of a fight. Others refuse to be beaten, and emerge from such trials seemingly stronger than before. In the 1980s, both Gaia and Lovelock were tested, and neither was found wanting.

The first sign of the heart trouble that would later threaten Lovelock's life came in September 1972, on a visit to the United States in connection with his work for NOAA on labelling air masses travelling across North America. Clutching a heavy bag and hurrying to catch a flight from Salt Lake City to Idaho Falls, he felt an odd pain in his lower chest, but 'it soon went away, and I thought it was just a result of running in a lopsided fashion while carrying the bag'. Later in the year, back in the USA, he was at a conference on CFCs in Andover, Maine, when the pain returned. This time it started to recur. As long as he stood still, he was fine; if he walked fifty metres, it returned. Lovelock knew enough medicine to suspect severe angina, caused by a partial blockage of one of the arteries in the heart, but he didn't really believe it. 'I was only 53, and fit as a flea. I couldn't believe it could happen to me.' So he did nothing, and travelled on the next day, as planned, to Boston to meet up with Lynn Margulis. By then, the symptoms could not be ignored, and when Lovelock explained the situation Lynn and her husband, Nicky, took him straight to the local hospital. After an ECG and X-rays, the hospital offered to admit him at once, citing high blood pressure, angina, and a 'very poor' ECG, any one of which required urgent attention. But Lovelock had no health insurance, and knew the cost of American medicine. He insisted on returning to England, armed with nothing more than a few trinitrin tablets, which have the effect of widening the arteries and improving the flow of blood, to relieve the worst symptoms.

The response of St Mary's Hospital in London was much more laid back than that of the American doctors; after the usual battery of tests, they simply told him to go home and consult his GP. After more tests at Salisbury Infirmary, Lovelock was prescribed more trinitrin for when the pain got bad and Aldomet for his high blood pressure, and told to take life easy. He also gave up cigarettes. But a month of the sedentary life was enough. He decided that exercise would be the best way to improve the functioning of his heart, whether or not one of the arteries was partly blocked, and started on a regime of hill walking on the nearby downs, popping trinitrin whenever the angina came on. 'Within a month, I began to feel much better. After six months, I was fitter than I'd ever been. Without trinitrin, the pain came on after I walked a hundred yards on level ground; with it, I could do anything.'

It was trinitrin that kept Lovelock going through the 1970s and the writing of his first book about Gaia. It was carelessness, he says, that nearly brought an end to his life on New Year's Day in 1982, a few months after the death of his mother, who had been living in Plymouth to be near to Jim and Helen. While driving the small tractor he used for odd jobs around the grounds of Coombe Mill, Lovelock failed to notice a patch of ice which made the tractor slip sideways and overturn on a slope, pinning him beneath the steering wheel with such force that the wheel was bent. Having extricated himself with great difficulty, and in considerable pain, he sought medical help, but was confronted by the total shut down that Britain experiences on a Bank Holiday. By then, Helen was using a golf cart to get around, and could offer little help. With neither of them physically able to drive, short of calling an ambulance the best he could do was telephone a locum doctor, who offered his 62-year-old patient the classic advice of taking a few aspirin and waiting to see if things got better overnight. The next day, the worst of the pain had indeed passed, and Lovelock carried on, although feeling tired and being less active for a month or so. Much later he learned that when he had been crushed his left kidney had been so badly damaged that it stopped working permanently. But 'there is a lot of redundancy in the human body, and we don't really need two kidneys. Only having one has never bothered me.' This was, though, just the beginning of the medical problems he would endure in 1982.

In April, Lovelock was attending a meeting of the Marine Biological Association in Plymouth. On his return to the hotel after a particularly rich meal with friends he had to climb two flights of stairs to the bedroom, and was hit by the worst attack of angina yet, experiencing for the first time the 'crushing pain' that is a classic symptom of a heart attack. The trinitrin still did its trick, but it was clear that something bad had happened to his heart. Lovelock's response was to go on a low-fat diet, and to buy a bicycle on which he could ride around the lanes near Coombe Mill. With the symptoms eased, Lovelock saw no reason to cancel his imminent visit to the USA to visit Hewlett Packard, NOAA and Lynn Margulis. 'Being a freelance means you can't take time off sick without a drop in income. Besides, I've always hated the idea of doing nothing.' He also, with mixed feelings, fitted in a visit to Dartington Hall, sixty miles from Coombe Mill, to give a talk about Gaia. Dartington Hall is a home of the Green movement, and by 1982 Lovelock had realized that the association of Gaia with New Agers and the more hippy kind of Greens was doing it no good in scientific circles. Whatever his reservations, it turned out to be an enjoyable day out. He met Jonathon Porritt, who Lovelock came to regard as the sane voice of the Green movement, someone he can debate issues with even though they often disagree with one another. Visits to Dartington became an escape during 'one of the more trying years of my life'; although Lovelock didn't always see eye-to-eye with the Greens, he valued the opportunity to debate the issues with people he regarded as friends.

The tribulations, described in graphic detail in his book *Homage to Gaia*, began almost as soon as Lovelock returned from his American trip, which passed off without a hitch. Applying his now familiar philosophy of treating a dodgy heart with exercise, he stepped up his regime of cycling and walking around the lanes. The walking was more demanding, round trips of about five miles taking just over an hour and involving a climb of about a hundred metres. They required taking a trinitrin pill every half mile or so, and certainly affected his heart. 'Several times during May and June I fainted, waking up a few minutes later to find myself lying on the road.' Fortunately, there was little traffic and nobody ever came across Lovelock lying by the roadside. The faints must have occurred when the blood supply to his heart was temporarily reduced virtually to nothing. Although the angina became almost ever present and recurred even when Lovelock was resting, 'I lost weight and grew fitter.'

Even Lovelock realized that he couldn't go on like this, and in July he sought medical advice, only to be told that his ECG was fine and there was nothing to worry about. He knew the diagnosis was wrong, but carried on through the summer. Looking back, he jokes that the three residents at Coombe Mill added up to about one healthy person. Helen was by now almost unable to move on her own, and was increasingly concerned that Jim's exercise regime would kill him; their son John had epilepsy and other problems but was physically able. Together, John and Helen would look after the day-to-day running of the household and garden, while Jim kept working to bring in the money. He also acted as a 'surrogate physician' for Helen when doctors were not available and a crisis occurred. On one occasion, he remembers, there was a crisis at a weekend when Helen suffered a bad reaction to her medication, which Jim decided was causing a lack of potassium. With no one else to turn to, he gave her some potassium citrate, from among the chemicals in his lab, dissolved in orange juice. 'The effects were like magic. Within a few minutes she was her old self again.' But had his diagnosis been wrong, he could have killed her. 'I learned then why physicians seldom treat their own families.'

Jim's escape during these troubled months came when he visited Dartington Hall. But Helen had no escape, and Lovelock let the contact fade away in order to spend more time with her. Partly because of Helen's illness, since moving to Coombe Mill the couple had been 'like two friends sharing a house', he says, bound together by four decades of shared memories and happy in a quiet way. 'I would never have known what we missed out on if I hadn't met [my second wife] Sandy.' Nevertheless, Lynn Margulis, who knew both Jim and Helen well, says that he was 'crazy with grief' when Helen became ill, and distressed that he could do nothing to halt the progress of her disease.

She confirms that Helen was 'a wonderful woman, intelligent, capable and supportive', who provided just the kind of companion Jim needed for many years. But 'his bond with Sandy is particularly strong. Perhaps now his bonds with old colleagues have become weaker as he and Sandy are such a close unit.'

In the autumn of 1982, Jim decided once again to seek medical advice about his heart condition, but before he could do so he received an invitation to visit Japan to talk about Gaia. In spite of everything going on in his personal life, he decided that this was too good an opportunity to miss, and he turned it into a round the world trip, travelling out via the United States to make his regular calls there. It was only when he got back to England that, following advice from his medical friends, he consulted the cardiologist Douglas Chamberlain, in Brighton. When told that the blood supply to his heart was so deficient that he needed immediate surgery, Lovelock's first reaction was relief. He had been sure in his own mind that that was the case, and having someone with expert knowledge agree and decide to do something about it 'took all of the year's anxieties off my shoulders'.

An angiogram taken a few days later at King's College Hospital in London showed why Lovelock wasn't already dead. Although he had a complete blockage of the left coronary artery, which ought to have been fatal, there was a small artery connecting the right coronary to the left, below the blockage, and blood had been flowing through this artery to partially compensate for the block. As one of the doctors put it, Lovelock had 'built his own bypass'. His life hanging almost literally by a thread, he was booked in for surgery a week later. Characteristically, although advised to stay in hospital to await the operation he insisted on going home in the meantime to make arrangements for Helen and John. 'I regarded the whole thing as a grand adventure. It never occurred to me anything might go wrong.' The bypass operation, carried out on 21 December 1982, was indeed a complete success; but the aftermath resulted in pain and misery that persists to the present day.

The problem was that the catheter inserted through the penis to drain his bladder during the long operation had not been sterilized properly, and came out 'covered in blood and dotted with shreds of adhering tissue'. The damage to Lovelock's urethra required extensive reconstructive microsurgery. 'I've lost count, but it's at least forty separate operations.' The ultimate cause of the sterilization problem was an industrial dispute which had led to a 'go slow' and work to rule by some of the hospital employees; Lovelock remains furious that anyone should put lives at risk by such an action, but holds no ill will towards the hospital or the National Health Service. If anything, his experiences over the next twenty-five years reinforced his belief in the desirability of a free medical service available to all. 'Just think how much the heart surgery would have cost me if it had happened in America; of course I never thought of suing the hospital or anything like that. They literally gave me a new lease of life; how could I be so ungrateful as to sue them?' Nevertheless, as the years have passed the problems have increased, especially in the form of repeated urinary tract infections, and have caused 'incredible trouble' for Jim. 'They still treat the infections with antibiotics, but as you get older the medication affects your brain more. I can't think clearly while it's going on.'

Lovelock returned to Coombe Mill on New Year's Day 1983, exactly a year after the tractor accident, and was able to complete his five-mile walk with ease, but the urethra problems soon began to make themselves felt. In between visits to hospital, he tried to carry on with his work, but 'life from February until June 1983 was a nightmare of pain and despair.' The first real help came when another of Lovelock's colleagues from the old Mill Hill days, David Pegg, recommended Michael Bishop, a surgeon based at City Hospital in Nottingham, who made the first partially successful repair and gave him eight weeks' relief. He then began a long programme of operations performed by Mr P. I. (Paddy) O'Boyle at Taunton to reconstruct his urethra over a period of several weeks in the autumn of 1983. A series of bladder and kidney infections followed, as a result of which the damage to Lovelock's left kidney was revealed, and in the end it had to be removed.

In among all this medical activity, somehow Lovelock managed to squeeze in a visit to Vienna to talk about Gaia, and an interview with a BBC crew who visited him at Coombe Mill while making a *Horizon* programme on the subject. It was a sign of things to come.

Towards the end of the year, the bout of repeated operations was over, and Lovelock was left with a functioning penis and a working urethra, the chief remaining drawback being that every couple of weeks he has to dilate it by inserting a catheter. 'But that isn't so bad. Rousseau had the same problem, but he had to use willow twigs, not a piece of smooth plastic tubing.' He was ready to get back to more intensive work on Gaia, responding to the criticisms triggered by his first book. Serendipitously, Lynn Margulis arrived on a visit, late in 1983, with 'a wonderful gift'. On behalf of the American Commonwealth Fund Book Program, Lewis Thomas, then Chairman of the Fund, invited him to write another book about Gaia, for which the Fund offered a grant of $50,000, to be repaid from the book's royalties. This was enough for Lovelock to concentrate on the book for the next couple of years. The book would be the ideal way to respond to his critics, and the timing was perfect. As he recovered full fitness, he set to work without distractions; but in fact what he now regards as his best idea ever had come to him during his difficult years, and had already been published.

The book that Lovelock wrote for the Commonwealth Fund was *The Ages of Gaia*, which first appeared in 1988. His first book was deliberately written for a wide audience to spread the idea of Gaia and stimulate debate; *Ages* was written for a more scientifically informed audience, to make the case that Gaia is indeed a theory, not merely a hypothesis. Lovelock emphasizes that because of the way the book was commissioned by the Commonwealth Fund, it was 'peer reviewed' by several eminent scientists, including Lewis Thomas and H. D. Holland, then Professor of Geology at Harvard University (who 'made more than 250 comments!'), and revised in the light of their criticism before publication. This is important to him, because he has been criticized for not publishing all of his ideas in peer-reviewed journals; but he did publish them in a peer-reviewed book. 'They were tough reviewers,' he says, 'and many changes were made. *Ages* is my most important book. It's the one I'd most like fellow scientists, in particular, to read.' Picking up from where we left off in Chapter 7, we will highlight just two key points from the book: the way Gaia regulates temperature, and the way Gaia controls the chemical environment.

The first criticism that Lovelock set out to answer was Dawkins' point about how Gaia 'could evolve her global adaptations by the ordinary processes of Darwinian selection acting within the one planet'. Because, as Margulis had pointed out, everything on Earth is connected to everything else, it was an impossible puzzle to tackle in the real world. Dawkins had written, 'I very much doubt that a model of such a selection process could be made to work.' Lovelock was determined to prove him wrong. The sheer complexity of the interactions between living things and between living things and their physical environment meant that a drastic simplification was required – a scientific model, a set of equations, which could be used to highlight one aspect of how the world works, without being a representation of the actual world we live in. Lovelock's experience in systems design helped him to come up with this kind of solution, a mathematical model that any physicist or engineer would feel comfortable with, which ultimately played a key part in establishing the credentials of Gaia as a true theory. Ironically, though, it was an approach which did not fit in with Margulis's style of 'hands on' biology involving the investigation of real living systems, and from the early 1980s onward the two people who had done so much to get the Gaia idea off the ground in the 1970s began to follow different, but complementary, paths in their further development of the idea.

The value of mathematical models is that they enable you to pick out one aspect of a system and study it – and to study things that are physically out of reach. A classic example comes from astrophysics, where scientists on Earth want to know what goes on inside a star. They use the best data they can obtain about things like the way atomic nuclei interact with one another, which they get from studies using particle accelerators here on Earth. Then they add in the basic laws of physics – things like Newton's law of gravity, and the laws relating temperature and pressure in a gas – and develop a set of equations to describe what is going on in the middle of a star. This model makes predictions about what the star should look like – for example, how hot the surface of a star with the same mass as our Sun should be. These predictions can then be compared with observations of real stars to see how good they are. If the predictions match the observations (which they do with great accuracy for today's astrophysical models) then we know that the model is a good one, and is telling us something real about the way stars work. And *then* we can use it to calculate other things, such as the way the temperature of the Sun has changed over geological time.

For months, Lovelock used this kind of approach to chip away at the problem of how the temperature of Gaia could be regulated by the behaviour of organisms acting solely in their own self-interest, turning over various ideas in his mind until, at the end of 1981, he composed the computer program for Daisyworld. It was first given a public airing at a conference in Amsterdam in 1982 and published in the journal *Tellus* the following year, while Lovelock was struggling with his heart problems, in a paper co-written with Andrew Watson, of Reading University. Lovelock had been offered the purely honorary title of Visiting Professor at Reading University in 1965 by his friend Peter Fellgett, who heard about the difficulty Jim was experiencing getting scientific papers published from a private address; the arrangement lasted until he was forced to 'retire' by university regulations at the age of 65, in 1984. There were no duties attached to the title, no money changed hands, and he seldom even visited Reading, but one fruitful outcome of the relationship was that in the second half of the 1970s Lovelock acted as the supervisor for Watson when he was studying for his PhD. Watson provided some of the key input for this stage in the development of Gaia theory, and is now a professor in his own right, at the University of East Anglia. He was an ideal collaborator, because Lovelock needed a mathematician who could write a paper in language that mathematicians would accept, and Watson is a whizz at maths.[1]

Watson's path to his present post was not always an easy one, and he is sure that his early career suffered through his association with Gaia. 'It is not,' he says, 'an idea that is embraced by many in the old-established departments in the old-established universities on either side of the Atlantic. They dislike the resonances in the name.' On the other hand, as time passed Gaia became an advantage. 'Most of my best students have found their way to me after reading about

1. Watson recalls his first visit to the Lovelocks, at Bowerchalke in 1976, to discuss the possibility of Jim being his supervisor. 'The whole Lovelock family seemed to inhabit a universe slightly tilted relative to the rest of the world . . . we spent a great afternoon talking about science. [Jim] was chock full of ideas that I'd never heard of, and excitement about science. He was a great PhD supervisor.'

Gaia.' And the particular feature of Gaia theory that draws them to Watson is still Daisyworld.

In its simplest form, Daisyworld is very simple indeed. It is an imaginary planet in the same orbit around the Sun as the Earth, but with a very simple ecology. The model deals only with one feature of the environment, the temperature, and all of life on the planet is reduced to a single kind, different varieties of daisy. Such an extreme simplification, designed to highlight one aspect of reality rather than a complex system like the models used by weather forecasters, is sometimes called a 'toy' model. But toy models can provide profound insights into how nature works. In the Lovelock and Watson version of Daisyworld, daisies cannot grow if it is too cool (below 5°C) or too hot (above 40°C). They do best at a temperature close to 20°C. The planet is warmed by radiant energy arriving at its surface from the Sun, and this is balanced by the heat that the planet radiates out into space in the infrared. The influence of clouds and the greenhouse effect is ignored. Apart from the amount of heat arriving from the Sun, the only thing that affects the temperature at the surface of Daisyworld is how shiny the surface is – its albedo. A surface that reflects no radiation at all has an albedo of 0; one that reflects all the incoming illumination has an albedo of 1. An albedo of 0.3, for example, means that it reflects 30 per cent of the incoming energy and absorbs the rest.

The daisies on Daisyworld have a variety of colours, ranging from dark daisies with an albedo of 0.2 to light ones with an albedo of 0.7 – in the very earliest model, there were just two kinds of daisy, light and dark. When the world was young and the Sun was cool, the only region of Daisyworld where the temperature was above 5°C was around the equator, so that was where daisies started to grow. But if there was an even mixture of dark and light daisies to start with, the darker daisies would do best. Because they absorbed more heat from the Sun, they would warm their surroundings above 5°C and flourish, producing seeds to spread the genes responsible for their dark colour in the next generation. Light coloured daisies would start to grow, but would reflect away the incoming solar energy and cool their surroundings below 5°C, so they would die, except where they happened to be in the midst of a patch of dark daisies. There would be two long-term effects. In future generations, there would be more dark daisies than light ones in the population; and the daisies would spread out from the equator as they warmed their surroundings, exerting a bigger influence on the heat balance of their planet.

It is easy to put the numbers into a simple computer simulation to see what happens as the generations pass and to include the effect of steadily increasing energy output from the Sun. When the solar output is low, dark daisies flourish and absorb heat, raising the temperature close to the 20°C optimum. This more or less steady temperature is maintained for a long time, thanks to a changing balance between dark and light daisies, even after the solar output has reached the point where the temperature at the surface of Daisyworld ought to exceed 20°C. From then on, light-coloured daisies dominate. As the temperature rises, dark daisies are less favoured and light daisies are more favoured, until we reach a stage where the entire planet is covered by light-coloured daisies reflecting as much heat as possible away into space. At this point there is a crisis as the solar energy becomes too hot to handle in this way; all the daisies die and the temperature suddenly soars. But the daisies keep the temperature of the planet essentially stable as the Sun's output increases from about 60 per cent of its present value to about 140 per cent of its present value. It jumps quickly from an initial cold state to something close to the optimum temperature, and stays there for billions of years before eventually jumping catastrophically away from the equilibrium.

The beauty of Daisyworld as a system lies in a combination of positive and negative feedback. If the world is cold, dark daisies do better and better in a positive feedback until the temperature reaches 20°C; but then, dark daisies warm their surroundings above the critical temperature and start to wilt, reducing the temperature so that the hotter it gets, the less they are favoured, in a negative feedback. But the absolutely crucial point is that at every stage every single daisy is acting in accordance with Dawkins' doctrine of the selfish gene. By acting solely in their own self-interest,[1] the daisies maintain a

1. Of course, genes are not literally selfish; this is Dawkins' metaphor. One of Lovelock's responses to the criticism by Doolittle and Dawkins is that 'Gaia is alive in the same way that genes are selfish.'

comfortable environment not just for themselves but even for their evolutionary rivals. The temperature on Daisyworld is regulated without any need for foresight or planning by the daisies.

No wonder Lovelock calls Daisyworld 'my proudest scientific achievement'. It completely answers the doubts of Doolittle and Dawkins that Gaia could evolve her global adaptations by the ordinary processes of Darwinian selection, and offers a model of how such a selection process could be made to work. It doesn't matter that the model is obviously an oversimplification; most models are. The point is that it can be made to work. 'I'm deeply grateful to Doolittle and Dawkins for pointing out flaws in my original argument and leading me in this direction,' says Lovelock. 'It represents the real spirit of constructive scientific criticism.'

It's important to appreciate that Daisyworld is not intended to be a representation of how the real world regulates temperature. It is a simple model to show that it is possible for Darwinian evolution by natural selection alone to come up with a way of keeping the temperature stable, even though every single organism involved is acting 'selfishly'. Once you accept that it is *possible* for evolution to do this, it follows almost inevitably that nature, or Gaia, will find a way to do it, even if the interactions involved are far more complex than in our models. Lovelock was not trying to model the way the real world works; he was showing that Dawkins was wrong to doubt that 'a model of such a selection process could be made to work'. By and large, physicists love the idea of Daisyworld, while biologists are sometimes uneasy with it. But that hasn't stopped it being developed into a sophisticated tool for illustrating how living things can interact to produce a whole that is greater than the sum of its parts.

Of course, in the early 1980s Daisyworld was further criticized for being too simple. In the same spirit of responding to constructive criticism, over the decades since 1982 it has been refined and improved in the light of those objections, passing every test, to give even better insight into the way the world works. Lovelock and other enthusiasts for Daisyworld introduced variations on the theme in which there are dozens of different kinds of daisy, each a different shade of grey, competing for space on the planet. Lovelock invented a version of Daisyworld in which the daisies are eaten by 'rabbits' and the rabbits are eaten by 'foxes', all simulated in a computer model. He then played with this to see what would happen to the populations of daisies, rabbits and foxes if the world was hit by a catastrophe such as the sudden death of 30 per cent of all the daisies. After short-lived hiccups, the system always settles down back at the optimum temperature – but he noticed that the hiccups, involving swings of temperature above and below the optimum level, are always bigger if a crisis strikes when the system is near to the temperature limit above which daisies cannot survive. This has ominous implications in the light of present forecasts of how the anthropogenic greenhouse effect will increase global warming, which we discuss later.

Andrew Watson emphasizes the continuing importance of Daisyworld and its value, not least as a teaching aid, in showing the basic point that regions of stability are separated by unstable regions, so that the climate system can flip from one stable state to another, very different state. 'These properties have been borne out by many subsequent climate models, but Daisyworld emphasizes that the biology of the system has a key role and that when the climate state changes drastically the biology also changes drastically – the two drive each other.'

Mark Staley, of the University of Guelph, Ontario, is among the researchers who have used variations on the Daisyworld theme to study evolution at work; in a paper published in 2002 he summed up his findings:

> [The] co-evolving dynamical process eventually leads to the convergence of equilibrium and optimal conditions . . . Sensitivity analysis of the Daisyworld model suggests that in stable ecosystems, the convergence of equilibrium and optimal conditions is inevitable, provided there are no externally driven shocks to the system. The end result may appear to be the product of a cooperative venture, but is in fact the outcome of Darwinian selection acting upon 'selfish' organisms.

What we are doing now, Lovelock points out, is providing an externally driven shock to the Earth System by injecting large quantities of greenhouse gases into the atmosphere.

There are variations on the Daisyworld theme where the daisies mutate, as real organisms do, so that their offspring may be darker or lighter than their 'parents', and versions with neutral-coloured daisies that 'cheat' by benefiting from the effort other daisies put in to making pigment. There are versions in which catastrophes and plagues of various kinds affect the system, which test how resilient this kind of system is. There is even a computer game, SimEarth (developed by the people who brought you SimCity) which includes a version of Daisyworld and pays a royalty to Lovelock's charity, Gaia, which he set up to further work on the theory. This all goes to show that it is indeed possible to imagine ways in which evolution by natural selection can produce a system like Gaia. But there is no getting away from the fact that Daisyworld is just a model, and although it may be telling us important things about how the real world can work, it is not the real world. The other dramatic step for Gaia in the 1980s was the discovery of a real world system which is seen to be operating on Gaian principles. Even better, as with all good scientific theories the theory made a prediction and the real world was then discovered to be operating in line with that prediction. 'Daisyworld was my greatest invention, but the sulphur cycle was my greatest discovery.'

As we have seen, Lovelock was already interested in the sulphur cycle and the role of dimethyl sulphide in the early 1970s, at the time of the *Shackleton* voyage. One fortunate outcome of that voyage was that Lovelock made friends with a young German researcher, Hans Greese, who was on board taking measurements of carbon monoxide. Greese arranged for Lovelock to visit the Max Planck Institute of Atmospheric Science, in Mainz, to talk about his work, and as a result of that visit he was invited to sail on a cruise of the German research ship *Meteor* in 1973, from Hamburg to the Caribbean. His original intention was to carry out more monitoring of CFCs, but the *Meteor*, as well as being a much larger and more comfortably appointed ship than the *Shackleton*, had an enclosed air-conditioning system. It turned out that many of the scientists on board were using CFCs and related compounds in their work (for example, as solvents), while most of the ship's company used deodorants and shaving foam in spray cans powered by CFCs. As a result, the air in and around the *Meteor* had 'the highest concentration of halocarbons in any air I have ever measured', and it was impossible to pick out the traces of CFCs from the background atmosphere.

Undaunted, Lovelock found something else to occupy his time. He noticed that after he had been standing on deck for a while both his skin and his shirt picked up a distinctive odour, similar to that of chlorine, that he recognized as due to a pollutant called peroxyacetyl nitrate, or PAN. The odd thing was, PAN is usually associated with man-made pollution, and had only been discovered as recently as 1956, as a component of Los Angeles smog. What was it doing out in the Atlantic? His curiosity piqued, Lovelock began monitoring the concentration of PAN in the air, and to his surprise found that it increased, rather than decreased, the further they got from land. Far out in the Atlantic, on a calm day when the *Meteor* had stopped to make other observations, he persuaded the Captain to allow him to travel a couple of kilometres away from the ship in an inflatable to obtain samples of absolutely clean air; the PAN was still present.

His studies soon showed that the amount of PAN in the air varied with the time of day. There was hardly any present at dawn or dusk, but a peak occurred around midday. Clearly, it was being made by photochemical reactions involving sunlight, and destroyed by other reactions as the Sun went down. So – what was it being made from? It turned out that the most likely candidate was methyl peroxynitrate, produced by the oxidation of methane in photochemical reactions also involving nitrogen oxide, and that this was then reacting further in the glass syringes used by Lovelock to collect air samples (or on his skin and shirt) to make PAN. In itself, this was not a major discovery; but it was an eye opener to discover that natural systems had ways of manufacturing the same pollutants found in smog, and to see how active the methyl radical, as it is known, is even in mid-ocean. Since methane comes from life processes, the discovery also offered another glimpse of the links between what were traditionally regarded as the separate 'living' and 'non-living' components of Planet Earth.

Because of his interest in marine organisms and their influence on the environment, in 1981 Lovelock had been invited to join the Council of the Marine Biological Association (MBA), based at Plymouth. Although this involved him in some tedious committee work, it also opened up new contacts in the field of oceanic chemistry, and gave him the opportunity to make two more sea voyages, on the ships *Challenger* and *Sir Frederick Russell*. He served as President of the

MBA from 1986 until he resigned in 1990, after fighting to retain a measure of autonomy for the Plymouth laboratory when it was merged with the nearby Institute for Marine Environmental Research as part of 'a mindless piece of bureaucracy that changed the MBA from an excellent independent laboratory into just another civil service outpost where good science came second to administrative convenience'.

In the early 1980s, although battling with ill health, Lovelock also made another important conceptual advance with Gaia theory. Although even then a few scientists were already beginning to be concerned about the implications of a buildup of carbon dioxide in the atmosphere as a result of human activities, Lovelock knew that there was another 'carbon dioxide problem'. Compared with Venus and Mars, the Earth's atmosphere has a remarkably small concentration of carbon dioxide, even though the gas is constantly pouring out of the Earth's interior from volcanoes. How is the extra carbon dioxide being absorbed? The conventional wisdom at that time had it that processes of rock weathering involving only physics and inorganic chemistry could do the trick. In the simplest form of this model, carbon dioxide in the air dissolves in water droplets that become rain, and when the rain falls onto rocks which contain silicate, such as basalt, there is a reaction which makes carbonate rocks like limestone and decreases the amount of carbon dioxide available to get back into the air. One obvious attraction of the model was that it seemed to offer a way to explain the reduction of carbon dioxide in the air gradually over billions of years, reducing the greenhouse effect as the Sun warmed and resolving the 'faint young Sun paradox'. The snag was, the process was nowhere near efficient enough.

Venus, which is a planet very similar in size to the Earth and, as far as we can tell, made of very similar material, has about 300,000 times as much carbon dioxide in its atmosphere as there is in the Earth's atmosphere. In round numbers, this does indeed match the amount of carbon dioxide stored in the rocks of the Earth, itself a strong hint that the two planets started out in much the same way. But the process of rock weathering we have described cannot alone account for all of the carbon dioxide drawdown that has taken place. Before human activities started increasing the proportion of carbon dioxide in the air, it was less than 270 parts per million; even now, it has not quite reached 400 parts per million. But the inorganic chemical and physical processes of carbon dioxide drawdown, operating over the entire history of the Earth, should have left about a hundred times as much in the air. What process could be more efficient than physics and inorganic chemistry at removing carbon dioxide from the atmosphere? To Lovelock the answer was obvious – life.

There are two ways in which life changes the scenario. First, living creatures such as plankton absorb carbon dioxide and use it to build their carbonate shells, which fall to the seabed when the creatures die and become part of carbonate rock layers like the famous white cliffs of Dover. But this alone is not enough to do the job. The marine creatures need a steady supply of dissolved carbon dioxide, and the gas doesn't simply dissolve from the air into sea water in large quantities. The second part of the contribution of life, Lovelock realized, is to increase the rate of rock weathering.

Living organisms act like a giant pump, taking carbon dioxide out of the air and carrying it deep into the soil. Some carbon dioxide escapes from the roots of a plant during its lifetime, and when the plant dies the roots are oxidized where they lie, releasing more carbon dioxide into the soil. There, the carbon dioxide is in intimate contact with calcium silicate, and reacts to make calcium bicarbonate and silicic acid, which dissolve in rainwater. Eventually, these compounds are carried in soluble form out to sea, where they are available for marine creatures to build their shells. When there is more carbon dioxide available, the marine creatures flourish and carbon dioxide is drawn down out of the air; when there is less carbon dioxide available, the marine creatures do not do so well, so the pump slows down. Over time, this maintains a balance which depends on the amount of carbon dioxide escaping from volcanoes. The ultimate source of that volcanic gas is carbonate rock which gets buried beneath the surface of the Earth, melts and is recycled by the processes of plate tectonics, closing the loop and forming a strong link between living and non-living processes.

Lovelock published papers drawing attention to these processes, in collaboration with his colleagues Michael Whitfield and Andrew Watson, showing that this process is at least tens of times more effective than inorganic weathering at drawing carbon dioxide from the air; by the end of the 1980s, other researchers had shown that weathering of basalt rock can be a thousand times faster if micro-organisms are present in the soil than if the rocks are sterile. Which explains why the Earth has not yet suffered a super-greenhouse effect – it is the presence of life that keeps the Earth cool enough for life. This was another triumph for Gaia theory; but by then, the big breakthrough had already been made.

The dimethyl sulphide (DMS) story involves methyl compounds, like the PAN story, and it involves a cycling of material between the oceans, atmosphere and land, like the rock weathering story. Although Lovelock had been interested in dimethyl sulphide since the early 1970s, and Peter Liss had also drawn attention to the link between DMS and life in the sea, the big breakthrough came in 1986, when Lovelock spent a month at the University of Washington, Seattle, as a Visiting Professor. There, he met an atmospheric scientist, Bob Charlson, who mentioned to him that one of the big puzzles in his own discipline was how clouds form over the oceans. Like most people, Lovelock had never imagined there was any such problem. Surely water simply evaporates from the warm sea into the air, being carried upward by convection and cooling until it gets cold enough to condense back into water drops? Charlson explained that this process would indeed work, but it would produce large drops of water that would fall straight back down out of a clear blue sky. Clouds are made from tiny droplets, so small that in effect they float in the air; and these tiny water droplets only form if there are very small particles, known as cloud condensation nuclei, in the air for them to form around. These condensation nuclei could be droplets of liquid, such as sulphuric acid from volcanoes, or very small solid particles like crystals of salt, or sulphate particles. These are the kind of tiny particles that are collectively known as 'aerosols', from which spray cans get their alternative name of aerosol sprays. There is no problem explaining where such things come from over the land, but they settle out of the air before it moves far offshore. Samples taken from the air far out over the Pacific Ocean showed, however, that there were indeed cloud condensation nuclei there, mostly in the form of sulphuric acid droplets and ammonium sulphate, but the atmospheric scientists had no idea where they came from.

Already intrigued, Lovelock became even more excited when Charlson told him why the puzzle was so important. If there were no clouds in the atmosphere, then, because clouds reflect back into space some of the incoming energy from the Sun, the average temperature of the Earth would be about 25°C, not the 15°C that it is today, and which seems to be so comfortable for life.[1] If life played a part in the production of cloud condensation nuclei, here was another component of the Gaian homeostasis system!

Both Lovelock and Charlson realized that DMS could provide the link between life in the oceans and clouds in the air, and together with two other colleagues, Andi Andreae and Steven Warren, in 1987 they published a paper in *Nature* suggesting how the link might work. 'This was one of the most important papers I have been associated with. We showed that DMSP [dimethyl sulphonium propionate] produced by microscopic life forms in the sea gets into the atmosphere, where it reacts to make particles that act as the seeds on which water droplets that make clouds can grow.' This was in itself an important, but uncontroversial, discovery. But they went further. If the cloud cover is reduced for some reason, then more sunlight would reach the surface of the sea, warming the planet but also stimulating photosynthesis and encouraging biological activity so that more DMS would be produced and the cloud cover would increase. But as the cloud cover increased, less sunlight would reach the sea, biological activity would slow down, and the growth in cloud cover would stop. If there were more cloud cover than today, the equivalent process would act in reverse to reduce the cloud cover and warm the world. It is another negative feedback mechanism which can help to maintain a stable temperature at the surface of the Earth and to restore stability when the system is disturbed. The Gaian link was at that time a speculation, but it made a clear prediction: if biological activity in the oceans increased for whatever reason, more DMS would be produced, there would be more of the compounds derived from DMS (such as methanesulphonic acid, or MSA) in the air, and the world would cool.

From the late 1980s onward, studies of layers of ice in cores drilled from the Antarctic were able to show how the composition of the air

1. Other things, such as the greenhouse effect, being equal.

changed, decade by decade and year by year, over tens of thousands of years. This is possible because each annual layer of snow that falls on the ice cap is compressed over the millennia to form a distinct layer, like the annual layers of wood laid down as growth rings in trees. From bubbles trapped in the ice, we know, for example, that during recent Ice Ages there has been less carbon dioxide in the atmosphere than there is today, providing a link between the greenhouse effect and Ice Ages. The cores also show that when the world was in the grip of ice the amount of MSA being deposited in Antarctica each year was between two and five times greater than today, showing a link between life in the sea, cloud cover, and cooling. The Gaian prediction had been borne out. Charlson and his colleagues calculated that the extra cloud cover implied by the measured amount of MSA found in Antarctic ice would have been sufficient for a cooling of the globe by 1 °C, slightly less than the cooling attributable to the reduced carbon dioxide concentration, but still a significant contribution to the Ice Age conditions. One implication of all this, Lovelock believes, is that Gaia may actually 'prefer' these cooler conditions to those we experience today. An Ice Age may be bad news for life on land at high latitudes, but it seems to be good for life in the sea and possibly for life in the tropics. The mere fact that carbon dioxide was taken out of the atmosphere and more MSA was produced shows that overall there was more biological activity going on during the recent Ice Ages than there is today.

But that is getting ahead of our story. What mattered in the mid to late 1980s was that Lovelock and Charlson had come up with a Gaian mechanism operating in the real world that made predictions which could be tested – the fact that it passed the test was almost a bonus. Together with Daisyworld, this marked the time that the idea of Gaia could be said to constitute a proper scientific theory, rather than a mere hypothesis.[1] Since then, the operation of the sulphur cycle has been the focus of a great deal of attention in order to find out how the different organisms involved obtain a 'selfish' benefit at the various

1. Lynn Margulis, who was careful to use the term 'hypothesis' in her early writings on Gaia, feels that it is now 'definitely a theory. A new theory has explanatory power and generates new observations and new work and enthusiasm.' And she still prefers the term Gaia to Earth System Science.

stages of a cycle which acts, overall, for the benefit of Gaia as a whole. 'We still don't fully understand it, but we've got some good ideas and it has encouraged a huge leap forward in the investigation of the links between different components of the planet.'

The link is important because of the role of sulphur as a trace element essential for life. It is a component of the amino acids from which proteins are made, and of vitamins. Plants obtain sulphur from compounds dissolved in water, and animals obtain sulphur from plants. On land, the sulphur compounds are mostly derived from mildly acidic rain, which incorporates sulphur from gases released by volcanoes and, significantly, from the breakdown of DMS produced by marine organisms. When land organisms die, sulphur gets recycled back into the oceans by water running off from the land. This runoff provides a steady supply of sulphur for the marine organisms, which they use for their own purposes before it gets released back into the atmosphere.

A key point is that at no stage do the marine organisms and land organisms act specifically for the benefit of each other; each kind of organism does what is best for itself, but they have evolved a link which acts to the benefit of both of them. In some ways, this is reminiscent of the way the slender curved beak of a particular species of humming bird has evolved to fit the exact shape of a particular kind of flower. The bird can only feed off one kind of flower, but this means it has no competitors (except for other members of its own species) for the food; the flower can only be fertilized by being visited by one kind of bird, but it is absolutely certain that those birds *must* visit the flower. It is a kind of co-evolution, an idea which has been developed in detail by Stephen Schneider, notably in his book *The Coevolution of Climate & Life*. Schneider does not fully espouse Gaia theory, but he is one of many researchers who have found that Lovelock's approach to Earth System Science raises fruitful questions about the way the living and non-living components of our planet interact.

It is easy to see why the processes which cycle sulphur between the oceans and the continents are a good thing for life in both places today; the obvious question is, how did such a cycle ever get started? Lovelock's answer, now widely accepted, is that the 'selfish' reason for organisms in the ocean making the effort to produce sulphur compounds is that the sea is too salty for them. 'The salt content of the oceans is uncomfortably close to the upper limit that life can tolerate.' His work on freezing and thawing living cells in the 1950s helped him to see what was going on. Cells like those of microscopic algae have to be able to allow the passage of water and other substances essential for life across the cell membrane which protects them from the outside world, but the unwelcome salts of the ocean at large have to be kept outside the cells – or rather, the less salty water has to be kept inside the cells. Curiously, if the cell contained only pure water and it was immersed in brine, rather than brine leaking in to the cell the pure water would leak out, as if nature were attempting to dilute the overly salty solution outside. This is another result of thermodynamics at work, an example of entropy increasing. What matters is that to maintain their own integrity the cells need to manufacture benign, neutral salts that they can keep inside themselves, balancing the osmotic pressure that would otherwise destroy them. Dimethyl sulphonium propionate (DMSP) is very good for the job, and is manufactured in many microscopic organisms in the ocean and seaweeds. When the organisms die, or when they are broken up in the act of being eaten by other organisms that graze on them, the sulphur escapes into the atmosphere in the form of DMS. This acts for the benefit of life on land, even though this is in no sense the 'intention' of the marine organisms. Tyler Volk, of New York University, put it graphically in his book *Gaia's Body*: 'Neither biogenic DMS nor a biotic enhancement of weathering evolved *because* they cooled climate, and yet their existence perpetrated free Gaian effects that profoundly link all life.' The discovery that DMS is indeed a key component of the natural environment was, says Lovelock, 'the first useful prediction from Gaia theory'. So useful, indeed, that it has established a new branch of scientific research – the first international symposium on DMSP was held in Mobile, Alabama, in 1995, and there have since been three more such meetings.

Since the 1980s, a great deal of work has improved our understanding of the interaction between algal growth, DMS, MSA and cloudiness over the oceans. It is clear that, because clouds reflect away incoming solar energy, without the DMS produced by marine organisms the world would be a hotter place – although the experts still argue about just how much hotter it would be. But it has also emerged that there may be local, as well as global, benefits for life in the production of DMS.

In the 1970s, researchers found very large concentrations of aerosol particles in the air above Australia's Great Barrier Reef. At that time, they guessed that the coral of the reef was responsible for the high aerosol count, but just how the mechanism worked was a complete mystery. It was only thirty years later, in 2005, that a team from Southern Cross University in Lismore, Australia, reported the discovery of large concentrations of DMS in and around the corals of the Great Barrier Reef; a kind of algal mucus exuded by the coral actually contains the highest concentration of DMS found in any organism. As a result, a layer rich in DMS forms at the surface of the water above the reef, where it gets picked up by the wind and carried up into the atmosphere. The emissions of DMS from the reef are not large on a global scale and have no measurable effect on the world's climate; but they are very significant locally, and affect the cloudiness of the atmosphere above the reef and the amount of solar energy penetrating to the reef.

The researchers are now carrying out further studies to find out what kind of Gaian feedbacks might be involved in maintaining conditions suitable for the growth of coral in the region. The Australian team has already found that the corals produce more DMS when the symbiotic algae that lives within the coral is stressed by high temperature or increased ultraviolet radiation. Their suspicion is that this leads to an increase in aerosol concentration and cloud cover, in a negative feedback, decreasing both temperature and UV to stabilize the situation. This would be a perfect test bed for studying one of the key Gaian mechanisms, if there were time to carry out the necessary observations.

Unfortunately, because of global warming much of the reef is already stressed and on the brink of dying. If a significant portion does die, the local feedback will no longer be able to compensate for the global effect, and there will be a decline in DMS production which will switch the pattern over to a *positive* feedback which will reduce cloud cover and allow more ultraviolet radiation through, an unhealthy situation that could rapidly finish the coral off. In itself, this is an important warning about what happens if the planet is pushed to extremes. Gaian feedbacks operate satisfactorily to maintain stability within certain limits of variation in things like temperature; but if they are pushed outside those limits stabilizing negative feedbacks may be replaced by destabilizing positive feedbacks that produce a sudden switch into a completely new climatic state. As we shall see, this is exactly what Lovelock expects to happen before the end of the twenty-first century. We can also see such an effect at work in the relatively recent past – when the Earth switched in and out of the latest Ice Age.

There is an overwhelming weight of evidence that the pattern of Ice Ages seen in the past million years or so is related to changes in the pattern of warmth received at different latitudes on Earth in different seasons, due to the slight changes in the tilt of the Earth and the exact shape of its orbit as it moves around the Sun that we mentioned in Chapter 1. These are called the Milankovitch cycles, after the astronomer and meteorologist who studied them in the early twentieth century. Although there is no change in the amount of heat received over the whole Earth over a whole year, the seasonal changes mean that sometimes the Northern Hemisphere has cold winters and hot summers, while at other times it has mild winters and cool summers. The rhythm of the Milankovitch cycles exactly matches the rhythm of Ice Ages revealed in the geological record, with full Ice Ages about a hundred thousand years long separated by Interglacials (like the conditions on Earth today) which last for some ten or fifteen thousand years. But the Milankovitch cycles do not have a big enough effect to produce the pattern of Ice Ages on their own; there must be feedbacks at work which strengthen the link.

We have already mentioned one of these – MSA in ice cores from Antarctica shows that more DMS was being produced when the Earth was in the grip of ice, so there must have been more cloud cover, reflecting away more of the incoming solar energy. The same ice cores reveal that there was less carbon dioxide in the air at that time, reducing the greenhouse effect. Both of these changes can be explained if more iron is available to 'fertilize' the upper layers of the ocean when the world is cooler. There are two ways this might happen. The first is the effect of dust blowing off the dry continents during an Ice Age and fertilizing the oceans. The marine organisms involved in determining the composition of the atmosphere and cloud cover need nutrients, and in particular they need iron, in order to grow. When water is locked up in ice, soil dries out and dust bearing those nutrients gets blown out to sea. The second mechanism, which Lovelock favours, is that when the top layer of the ocean is warm it acts as a lid on convection, and iron and other nutrients cannot get mixed up into it from below. But when it cools it no longer acts as a lid, and the upwelling nutrients encourage biological activity near the surface. Either or both mechanisms could do the trick.

So once the Milankovitch cycles tip the Earth towards an Ice Age, feedback completes the job. Once an Ice Age is established, it lasts until the Milankovitch cycles produce the maximum influence in favour of melting the ice. Then, as the ice starts to melt the soil is damped down and the supply of dust to the oceans declines, while the oceans warm at the top and convection is suppressed, so the feedback operates the other way, flipping the Earth into an Interglacial state. Either state is a possible stable situation for Gaia; but there is no stable intermediate state – it is either Ice Age or Interglacial, with no in-between. The question which concerns Lovelock is whether the Earth is about to flip into another, even hotter, stable state. 'The history of our planet suggests the existence of such comparatively hot stable states.'

By the end of the 1980s, Gaia was becoming respectable. In that decade, Lovelock also gained one of his most important and influential supporters, Sir Crispin Tickell, Britain's former Ambassador to the United Nations, who he met in New York. Tickell recalls being immediately impressed with Lovelock's 'unique approach to ideas about the connectedness of the physical and biological environments', and from the outset preferred to refer to 'Gaia theory', although he later, as a true diplomat, took care 'to mention Earth System Science in the same breath'. The end of the beginning of the Gaia story can be seen, with hindsight, to have occurred in 1988, the same year that *Ages of Gaia* was published, when Stephen Schneider and Penelope Boston organized the first international conference to debate what was then still called the Gaia hypothesis. It was held in San Diego, under the auspices of the American Geophysical Union (AGU), and is particularly noteworthy because Schneider, the chief moving force behind the conference, strongly disagreed with Lovelock, but felt as a good scientist that such an interesting and controversial idea deserved to be properly debated by scientists. The breakthrough work on the sulphur cycle was a key factor in persuading the AGU that Gaia was indeed respectable science and worthy of debate within their series of Chapman conferences; the proceedings were published, under the title *Scientists on Gaia*, in 1991.

Jim Lovelock's personal life also changed dramatically at the end of the 1980s. At the time of that Chapman conference, Helen Lovelock was in the last year of her life, and during the second half of 1988 Jim had the responsibility of nursing her through the final stages of her illness. It was, he says, 'a hideous experience'. Against the advice of doctors and family members, but in accordance with Helen's own wishes, she stayed at Coombe Mill in the comforting surroundings of home, rather than being moved to a nursing home. In the first week of November 1988 she recorded a kind of testimony in which she referred to her life as 'an endless fight to hold the line against encroaching disability . . . I needed someone like Jim who never seemed to know I wanted him and who expected me to fend for myself. I would have died long ago if I had not had to fight. I think that Jim is just like me in that way and has fought as hard.' This bears out the testimony of Lynn Margulis, who knew both Jim and Helen well and describes her as 'a fine and remarkable woman'.

In January 1989, Helen became ill with pneumonia, and had to be taken to hospital, where she died, surrounded by her family, on 4 February. She was buried at a spot she had chosen at Coombe Mill. As the decade drew to a close, it was a time for new beginnings, both for Gaia and for Lovelock.

9
New Beginnings

Life isn't always neat and tidy. The new phase of Lovelock's personal life had already begun by the time Helen died, as a direct result of the burgeoning interest in Gaia in the late 1980s. At the time, and ever since, Lovelock regarded the first Chapman conference on Gaia as something of a disaster. Somewhat naively, he had expected everything to fall into place, and all the scientists to be completely persuaded of the merits of his idea. Although such expectations were always unrealistic, his disappointment at the time is understandable, not least since by then he was in his sixty-ninth year, approaching the biblical three score and ten, and felt that he might not get another chance to make the breakthrough. But even today he looks back on the San Diego meeting as 'a waste of time', failing to appreciate just how important a step it was to get the idea debated in a mainstream, cross-disciplinary scientific forum. Returning from San Diego in the early spring of 1988, Lovelock responded to his disappointment by throwing himself into his next Gaian commitment.

The commitment was one that Lovelock had agreed to 'in a moment of weakness' the previous year. He had been asked to participate in a Global Forum meeting, to be held in Oxford in April 1988, where the state of the planet would be discussed. Other speakers at the eclectic gathering would include Carl Sagan, at that time making waves both in scientific circles and in the media with his ideas about nuclear winter, and Mother Theresa. Lovelock's contribution, in spite of all the effort he had put in to preparing it, rather got lost in the crowd. But the meeting changed his life.

One of the organizers of the meeting was a London-based American woman, Sandy Orchard, who Lovelock met briefly at the opening formalities. Their paths did not cross during the formal sessions, but at the social events during the course of the week he found himself increasingly attracted to her, although he saw no sign that she felt anything for him. It was only after the conference dinner, held in the splendid setting of Blenheim Palace, that their eyes met, 'literally across a crowded room', and they realized that the feeling was mutual. 'For the first time in my life, I had fallen deeply in love.' The conference ended on a Friday, but Sandy and Jim stayed on together overnight before heading their separate ways, having promised to meet again soon.

The affair continued through the summer, with Jim and Sandy meeting at every opportunity, including her joining him at a gathering of philosophers, writers and 'Green notables', as Lovelock describes them, in Italy. Their personal situations turned out to be remarkably similar – Sandy's husband, David, was terminally ill with cancer, and like Jim she was living in a friendly and caring relationship with her spouse. It was Sandy's support, as much as anything, that enabled Jim to cope with Helen's decline over the following months. He was careful to keep all knowledge of his relationship with Sandy from Helen – 'it was the only thing we didn't share' – but otherwise made no secret of it. This was characteristic of a man who many of his friends and colleagues described to us as being 'completely open and honest, almost to the point of naivety'. He finds it impossible to dissemble, and is completely baffled when people are not honest in their dealings with him. Inevitably, this can cause difficulties even for those closest to him. Jim's new life provoked criticism from his children; Christine and John 'were shocked, but understood', while Jane and Andrew, who lived further away and saw less of Jim, found what they saw as a betrayal of their mother harder to accept. But 'time has healed the wounds.'

David Orchard died in 1990, and Jim and Sandy were married in February 1991, when he was 71. The years since, he says, 'have been the happiest of my lifetime.' Sandy echoes the sentiment. They seem to have had the great good fortune of saving the best in life for last, and their serene and obviously happy partnership and enjoyment of each other's company have clearly been a strikingly positive influence on Lovelock at a time when interest in, and support for, Gaia has never been higher.

For the happiness is not just a result of the changes in Lovelock's personal life. At exactly the time all this was happening, global warming became established not just in the minds of scientists but also in public debate and among the more far-sighted politicians as a real threat to human civilization. And this made Gaia theory increasingly respectable, while opening doors to places Lovelock had never dreamed of making his voice heard – including No. 10 Downing Street.

The moment when the reality of the anthropogenic greenhouse effect and the threat of global warming became established beyond reasonable doubt can be pinpointed precisely, to 23 June 1988. This isn't said solely with the benefit of hindsight – we said as much at the time, in our book *Hothouse Earth*, published in 1990. Our conclusion was based on the way a series of studies of climate change had been published during the 1980s. The background to the story is that after a pronounced warming between about 1910 and 1940, the records of global temperature changes available in the late 1970s showed a levelling off, followed by a decline in temperatures from the 1950s into the early 1970s. Then, the pronounced warming trend resumed. These observations naturally led people to be cautious, in the 1970s, about making forecasts of further warming based on the increasing output of carbon dioxide into the atmosphere from human activities, even though the Keeling curve showed that half of this gas was staying in the atmosphere and building up year by year. Indeed, there was some concern in the early 1970s that the world might be heading for a new Ice Age. But it soon became clear that a large part of the cooling trend was caused by pollution of the atmosphere from industry and fossil-fuelled power stations. Like the aerosol particles from a volcanic eruption, soot, sulphates and other particles from industry were acting like a sunshield across the industrialized Northern Hemisphere, in particular over Europe and North America. Reid Bryson, of the University of Wisconsin, Madison, dubbed this the 'human volcano'. Because most climate observing stations were in the Northern Hemisphere, this local cooling trend had masked a continuing warming in the Southern Hemisphere. The warming in the north resumed with full force as governments in the industrialized nations introduced measures to curb aerosol pollution and clean up the atmosphere.

Today, we have better global observations (including those from satellites), improved coverage of the Southern Hemisphere, and longer records. We also have data from proxy records, including analyses of the isotope ratios in shells found in deep sea sediments, and analyses of ice cores from glaciers in both Northern and Southern Hemispheres. As a result, the temperature trends of the twentieth century (and before) can be accurately explained as resulting from a combination of the cooling influence of volcanic eruptions, the human volcano, a small contribution from effects associated with the Sun's changing cycle of activity, and the anthropogenic greenhouse effect. None of these effects alone, nor any two or three of them in combination, can explain the observations;[1] but together they fit the actual temperature trend closely. And greenhouse gas emissions now dominate the trend. But that had still to be appreciated when James Hansen, of the NASA Goddard Institute in New York, began analysing global temperature trends, including, crucially, one of the first studies to separate out the trends in different latitudes of the Earth, and in the Northern and Southern Hemispheres.

Hansen and his colleagues started out by analysing all of the available data for a full hundred-year interval, covering the decades from 1880 to 1980. Their analysis divided the world up into a grid of eighty boxes, forty in each hemisphere, with temperature trends calculated from observations from all the monitoring stations in each box, before the data from separate boxes were combined to give regional, hemispheric and global trends. The NASA researchers surprised themselves, as well as a lot of other people, when they found that by 1980 the warming trend in other parts of the world was already overwhelming the cooling trend perceived by observers in the industrialized north. In particular, southern latitudes warmed steadily by about 0.4°C between 1880 and 1980, with no peak in the 1940s nor any subsequent decline. This is literally the cleanest record of large-scale temperature trends over that period. When their results were published in the journal *Science* in August 1981, the NASA

1. It's worth emphasizing this, since from time to time enthusiasts for volcanic mechanisms of climate change, or fans of sunspot cycles, try to claim that their pet process alone will explain everything; it won't!

team concluded that 'there is a high probability of [global] warming in the 1980s.'

Most researchers were cautious about accepting the evidence, and some suggested that the records from the Southern Hemisphere, which depended on data from relatively few measuring stations, compared with the well-covered north, might be unreliable. Hansen's team responded to these criticisms, answering every point, in an article in *Science* in May 1983, commenting also on the eruption of the volcano El Chichón in South America in 1982 and suggesting that aerosol from that eruption might stem the warming temporarily, 'but, barring improbable further eruptions . . . significant warming is still likely in this decade.' By 1987, Hansen and his colleague Sergej Lebedeff were able to look back on the peak warmth of the early 1980s, just before and after that volcanic eruption, using the same eighty-box analysis to show that by that time the world was warming evenly. A very similar picture was described by researchers at the University of East Anglia, who found that after the aerosol from El Chichón cleared, 1987 was the warmest year since the start of reliable observations in 1858, with 1981 and 1983 tied in second place.

By then, the 1980s already stood out as something quite remarkable in the historical record of temperature changes.[1] This was what convinced many people that human-induced global warming was upon us. As we wrote in 1989 for publication in *Hothouse Earth* a year later:

> Whatever they tell us about the scientific side of the problem, though, the set of research papers published by the NASA group in the 1980s shows the pace with which the political side of the story developed. In 1981, it was possible to stand back and take a leisurely look at the record from 1880 to 1980. Comments on the paper, and the NASA team's response to criticism, were published nearly two years later, with no sense of urgency. In 1987, the figures were updated to 1985, chiefly for the neatness of adding another

1. The decade doesn't look so remarkable from the perspective of 2007. As global warming has continued, with a rise in global mean temperatures of 0.6°C between 1975 and 2005, more than in the hundred years from 1880 to 1980, the six warmest years on record, in order of increasing warmth, now stack up as 2006, 2004, 2003, 2002, 1998 and 2005. There is every likelihood the 2005 record will have been broken by the time you read this.

half-decade to the records; those figures were two years out of date by the time they appeared in print. But by early 1988, even one more year's worth of data justified another publication in April, just four months after the last 1987 measurements were made, pointing out the record-breaking warmth now being reached.

In the four months it took to get the 1987 data into print, the world had changed again. Just a few *weeks* later Hansen was appearing before the US Senate, informing them that the first five months of 1988 had been the warmest for any comparable period since records began, and that the greenhouse effect was upon us. The forecast made in 1981 had come true. At that hearing, on 23 June 1988, he said:

> The Earth is warmer in 1988 than at any time in the history of instrumental measurements . . . the global warming is now sufficiently large that we can ascribe with a high degree of confidence a cause and effect relationship to the greenhouse effect and . . . the greenhouse effect is already large enough to begin to affect the probability of occurrence of extreme events such as summer heat waves . . . heatwave/drought occurrences in the Southeast and Midwest United States may be more frequent in the next decade.

Those words will be twenty years old by the time this book is published, but effective action to tackle the problem has still barely begun. Such foot-dragging would have seemed unbelievable in the heady days when concern about global warming put Gaian thinking in a new perspective and made Lovelock a sought-after adviser in some (but not all) high circles.[1]

His most significant contact was with the then Prime Minister, Margaret Thatcher. In October 1988, Lovelock was invited to a dinner at No. 10 Downing Street, for a gathering of the great and the good; after dinner, Mrs Thatcher made a point of singling him out for a discussion over coffee, taking a keen interest in his views on environmental issues; she was clearly au fait with his work. As a direct result

1. In some ways, it still does seem unbelievable. We wrote *Hothouse Earth* at the end of the 1980s, after many years covering climate change stories, precisely because the scientific case for global warming and the anthropogenic greenhouse effect had been made, and it seemed time to hand over to the politicians. We little imagined that two decades later we would return to the subject with some of those politicians still rejecting the scientific consensus.

of that meeting, in April 1989 he returned to Downing Street to participate in a seminar on climate change.

As an old socialist, Lovelock might seem to have little in common with Margaret Thatcher; but as a scientist he found that she spoke his language. Her background as a chemistry graduate from Oxford and her experience in industry made her one of the very few politicians who understood how science and scientists work, and it is greatly to her credit that as early as 27 September 1988, in a speech to the Royal Society, she predicted that environmental issues would 'usurp' the political agenda in the decade ahead. She was surely the only head of state who could have personally handled a scientific meeting like the environment seminar, and from the point of view of making political progress with the problem of climate change it is unfortunate that by this time she was nearing the end of her time in office. Lovelock enjoyed what he describes as 'the warmth of her patronage' during her remaining months as Prime Minister, including representing the UK in Brussels in 1989 at an EU meeting on environmental ethics.

Recognition for Lovelock's lifetime work came from many quarters in the 1990s. Honorary degrees from universities in Britain, America and Europe, the Amsterdam Prize for the Environment, the Volvo Environment Prize, Italy's Nonino Prize, and the one which, with all respect to the other awards, he describes as 'the most extraordinary', the Japanese Blue Planet Prize in 1997, his 78th year. Many of these awards came in the form of substantial cash prizes; all of them endowed Lovelock with personal prestige. In 2003, he received one of the highest civil awards in Britain when he was made a Companion of Honour; the Order is restricted to the Sovereign and just sixty-five ordinary members, chosen for 'services of national importance'. But what mattered to Lovelock most was that all of the fuss was really about Gaia. By the end of the 1990s, Gaia was respectable. Nothing showed this more clearly than the holding of a second AGU Chapman Conference on Gaia. This took place at Valencia in the year 2000, twelve years after the first such conference and neatly at the end of the millennium, a suitable moment for new beginnings. The conference itself was called 'Gaia 2000'; the updated proceedings were published in 2004 under the title *Scientists Debate Gaia*, and this time Lovelock himself was happy with the tone of the meeting and the level of scientific debate. It was very nearly everything he had (rather unrealistically) dreamed the 1988 conference might be, and the proceedings volume of the Valencia meeting is still one of the best guides to Gaia theory for anyone with a little scientific background.

The new beginning for Gaia studies can be put in perspective, like the earlier work, by looking at examples from theory and from observations of the real world. Before, we looked at the Daisyworld model and at the cycles involving DMS; this time, we return to the understanding of thermodynamics that underpins the whole story of life, and also look at, among other things, the way the Amazon rainforest regulates its immediate environment and influences conditions far around the world.

When energy is flowing through a system, interesting things happen at the edge of chaos. This can be seen most simply by thinking about a flow not of energy, but of water, something we have all experienced. Of course, the moving water is carrying energy, which is what matters, but it is easier to picture than the flow of solar energy across space. If a river is flowing slowly past a single rock sticking up out of the surface, the water divides neatly around the rock and joins up smoothly on the downstream side. If the river is in flood, the torrent of water rushing downstream smashes into the rock and forms a messy froth of irregular fluctuations with a choppy surface behind the rock – chaos. But at some intermediate rate of flow, just before the onset of chaos, the water flowing past the rock breaks up into little whirlpools or vortices, structures which detach from the rock and get carried off downstream, still swirling away within the overall flow of the river. An ordered structure has emerged spontaneously. Other vortices, on a much larger scale, occur in the atmosphere as hurricanes and tornadoes, also feeding off energy gradients. If you are caught in a hurricane it is hard to think of it as an ordered structure; but photographed from space its inherent structure is not only visible but beautiful. This kind of order exists on the edge of chaos, and life, feeding off the energy flow from the Sun, is an example of order at the edge of chaos. In their book *Into the Cool*, Eric Schneider and Dorion Sagan sum this up by saying that 'life must be regarded, at the deepest level, as a matter as much of energy transformation as of genetic replication,' and refer to life as a 'process', to emphasize its active nature.

We have gone into the physics behind all this in our book *Deep Simplicity*; all that matters here is to hold on to the idea that living systems result from processes which feed off energy, building local complexity well away from thermodynamic equilibrium, but always operating on the edge of chaos, where an increase in the rate of flow could spell disaster. The key is the 'energy gradient' off which complexity feeds. Too shallow, and there is no scope for complexity; too steep and you have chaos. The fundamental rule of thermodynamics is that 'nature abhors a gradient' and acts to reduce it. That's why if you drop an ice cube in a glass of water the ice cube warms up and melts while the surrounding water gets cooler. For Gaia, as we saw in Chapter 3, it means that life acts to convert high-quality energy into low-grade energy. This is the meaning of life.

These ideas go back at least to the time of Lotka, and were discussed by the physicist Erwin Schrödinger in 1944, in his book *What is Life?* As he put it, life 'feeds upon negative entropy', and '[the] most powerful supply of "negative entropy" [is] in the sunlight'. The ideas have been developed in the context of Gaia theory by Eric Schneider, of the Hawkwood Institute in Livingston, Montana, who has looked in detail at the way different ecosystems transform energy. Schneider initially trained as a marine geologist, and like Lovelock his background in the physical sciences gives him a different perspective on life from someone with a background solely in biological sciences. In particular, he is fascinated by the way energy flows through a system, including a biological system. The most intriguing examples, and the most relevant for humankind today, come from studying so-called successions, the different ecosystems that emerge in turn during the recovery following a disaster such as a forest fire. Cleared woodland that is left fallow in England, for example, always goes through the same stages in which grassland is succeeded by shrubs, shrubs are followed by conifers, and oak forest emerges after about 150 years. What is going on in energy-flow terms at each stage of the succession?

The common characteristic of all successions is that the early stages are dominated by organisms which have short lifetimes and reproduce quickly, but are not very good at holding on to raw materials and energy. Later stages in the succession have a higher energy flow through the living components, which means they 'degrade' the energy more effectively, hold on to the materials they need for longer as the waste products from one organism are used by other organisms, and are more complex. The more mature the system, the more efficient it is at recycling resources that go to waste in immature systems. In thermodynamic terms, a complex system is more effective at producing entropy than a simpler system. 'Global humanity,' says Schneider, today 'resembles a pioneer species colonizing a new niche; to achieve the global equivalent of successional maturity – to last in the biospheric long run – we will have to increase our connections with other species, and recycle our materials more adeptly through global biosystems of greater diversity and complexity.' And 'non-equilibrium thermodynamics connects life to non-living complex systems.'[1] It turns out that there are sound thermodynamic reasons for doing all the things we have been told are 'good for the planet'!

Schneider points to an experiment carried out in New Hampshire in the 1960s, where an area of forest was cleared and sprayed with herbicides. The runoff from the affected region was then compared over the following years with runoff from neighbouring regions that had been left intact. In the first year, the runoff from the cleared region was 39 per cent higher than in previous years, and in the second year 28 per cent higher. As Schneider puts it, the region 'leaked its most valuable resource, water' and along with the water the runoff from the cleared forest included five times more calcium and magnesium ions than in the runoff from uncut forest, sixteen times more potassium, and nearly three time as much sodium. This is direct proof that immature systems are leaky, while 'mature systems hold on to their materials, using energy to recycle their constituents in complex loops'. The whole system benefits from the way the system affects its environment.

Gaian effects can also be seen at work in both smaller and larger systems than forest catchment areas. One irresistible example shows that Charles Darwin himself had a view of the relationship between the living and non-living components of our planet that looks distinctly 'Gaian' – a delicious irony since the fiercest criticism of the Gaia idea in the 1970s came from evolutionary biologists who work

1. See *Scientists Debate Gaia* and *Into the Cool*.

within a framework of ideas known as 'neo-Darwinism'.[1] The essence of neo-Darwinism is that evolutionary selection operates at the level of the gene, and that different packages of genes (different organisms) are in competition with one another for resources within the ecological framework that surrounds them. What is specifically excluded from strict neo-Darwinism is the idea that life modifies the parameters of the physical environment for its own good. But Darwin saw such processes at work all around him, and spelled out some of the details in his last book, titled *The Formation of Vegetable Mould, Through the Action of Worms with Observations on Their Habits*, published in 1881, the year before he died. Usually referred to as 'the worm book', this is as good a read, and as insightful, as any of Darwin's books, summing up decades of study of what seem at first sight to be insignificant members of the web of life, but turn out to be major players. After reading the book, Joseph Hooker wrote to Darwin that 'I had always looked on the worms as amongst the most helpless and unintelligent members of creation; and am amazed to find that they have a domestic life and public duties!'

The 'duties' Hooker refers to are the ways in which earthworms modify their environment to make it a better place for life. Eileen Crist, of Virginia Tech, has studied the book from a Gaian perspective, and says that 'Darwin's investigation into how organisms shape their surroundings marks his worm book as pioneering science.' Rather than focusing on the way earthworms adapt to their surroundings, she points out, Darwin concentrates on the way earthworms transform their environment. In modern terminology, this is Gaian thinking, not neo-Darwinism. As they burrow through the soil, worms swallow earth, extract nutrients from it, and expel the rest along with their chemical secretions in the fine particles of 'casts' – known to Darwin's contemporaries as 'vegetable mould', although, as he pointed out, 'the term "animal mould" would be in some respects more appropriate.' Always a patient man, over many years Darwin studied the amount of earth processed in this way and brought up to the surface by measuring the speed with which stones gradually sank beneath the

1. 'Neo' because Darwin did not know about genes when he formulated the theory of evolution by natural selection.

surface of an unploughed field, calculating that about an inch of 'mould' was added to the surface every dozen years. He made enquiries about the activity of earthworms in his extensive correspondence with naturalists around the world, studied information about the way Roman remains and other antiquities were gradually buried by the action of earthworms, showed how worm casts are carried down slopes by the action of wind and weather, and concluded that earthworms are a significant geological force:

For every 100 yards in length in a valley with sides sloping as in the foregoing case, 480 cubic inches of damp earth, weighing above 23 pounds, will annually reach the bottom. Here a thick bed of alluvium will accumulate, ready to be washed away in the course of centuries, as the stream in the middle meanders side to side.

But, as Crist points out, Darwin also appreciated that the action of worms makes their surroundings better suited for life. Worms, he said,

mingle the whole intimately together, like a gardener who prepares fine soil for his choicest plants. In this state it is well fitted to retain moisture and to absorb all soluble substances, as well as for the process of nitrification. The bones of dead animals, the harder parts of insects, the shells of land-mollusks, leaves, twigs, &c., are before long all buried beneath the accumulated castings of worms, and are thus brought in a more or less decayed state within reach of the roots of plants.

And as plants flourish, worms thrive – the main source of food for earthworms comes from the leaves and other plant matter that falls to the ground.

Crist sums up Darwin's achievement by saying that 'in his last book, Darwin showed that worms have a significant impact on the appearance, chemical constitution, physical structure, geological shaping, and biological organization of the land.' No wonder Hooker was impressed! Crist puts all this in the context of Gaia – or, as she prefers, geophysiology:

Geophysiology emphasises that organisms play an important role in the creation of their environments. Organisms thereby play an important role in creating themselves, for the environmental conditions they contribute to forming subsequently exert selective pressures on them and their descendants: on straight evolutionary reasoning, it follows that organisms which create a favourable environment for themselves will tend to be selected for through consequent feedback effects from the environment. Darwin's understanding of the shaping force of earthworms on their surroundings resonates strongly with this view: *he did not take the land as the given background to which worms adapt, but saw it as a medium actively created and maintained, in large part, by these animals themselves.*[1]

It was, she says, a fine example of thinking 'outside the box'.

A modern example of the same kind of thinking involves a slight change in scale, to a very successful but more self-contained kind of living system that clearly operates on Gaian principles. The central theme of Lovelock's idea is the way homeostasis has maintained conditions suitable for life for billions of years on Earth – key aspects of this are the regulation of temperature within tolerable limits, and the regulation of the concentration of oxygen in the air. Another system which does a superb job of homeostatic regulation, in this case of the composition of its atmosphere, is a termite mound. Scott Turner, of the State University of New York, trained originally as a physiologist but became fascinated by the interplay between physiology, evolution, ecology and adaptation. He has made a particular study of how the so-called 'macrotermitines', which rely on the use of a kind of fungus which they 'farm' to partially pre-digest the cellulose on which they feed, regulate the atmosphere inside their mounds for the benefit of both themselves and the fungus.[2] The relationship probably grew up long ago when fungi got into the ancestral termites' food supply, but it has now evolved to the point where macrotermitines build structures called fungus combs within their nests, and provide the fungi with cellulose in the form of grass, bark, bits of wood and the like. The fungi digest the cellulose and convert it into sugars, providing high-energy food for the termites – another example, of

1. From *Scientists Debate Gaia*. Our emphasis.

2. Turner has shown that termite mounds do *not* regulate their temperature homeostatically, although this claim still appears even in the scientific literature, and widely in popular accounts.

course, of waste recycling in a mature ecosystem. The success of this partnership is shown by the fact that these termites are typically two or three times larger than termites which do all their digestion for themselves, and are very active. Their colonies may include a couple of million active workers, ten times more than the colonies of termites that digest their own cellulose. But what we are interested in here is how all these workers keep their homes comfortable.

Two million large termites need a large place to live. Their mounds can be several times the height of a man, and are a dominant feature of the landscape across much of the savanna of southern Africa. But this is not where the termites live. Their nest is a compact, dense mass about two metres across, at the base of the mound. The mound, which is riddled with a complicated network of tunnels and chambers, is the air-conditioning system that ensures that the termites and the fungus benefit from the optimum concentration of carbon dioxide in the atmosphere in the nest. At the most basic level the mound acts as a ventilation shaft for the nest. The bigger the colony, the greater its overall need for respiratory gases, and the bigger the mound the more effective it is at catching the wind and providing more ventilation. But the control of the nest's atmosphere is much more subtle than this, because the structure of the mound can be adapted by termites who respond to changes in the environmental conditions in the nest by sealing up some of the holes in the surface of the mound, making new holes, or building the mound higher. Turner has shown that the effect of all this activity is to maintain a high and stable concentration of carbon dioxide in the nest.[1] This seems to be a benefit for the fungus farms, not least because it prevents them being invaded by another species of fungus.

The particular fungus favoured by the termites is known as *Termitomyces*, and is a good partner for them because it grows relatively slowly and leaves plenty of 'waste' for the termites to eat. Another fungus found in the soil of the regions where the termites live, *Xylaria*, grows more rapidly and digests the cellulose more completely; if it

1. He also told us that his latest work indicates that 'these termites are avid regulators of moisture in the nest,' but details of this work had not been published at the time this book was completed.

invades the fungus farms it is a disaster both for *Termitomyces* and for the termites. But *Xylaria* prefers a lower concentration of carbon dioxide and *Termitomyces* prefers a higher concentration of carbon dioxide. So both sides in the termite/fungus partnership gain from the relationship. *Termitomyces* gets its food supply provided by the termites, and is protected from its competitor; the termites get a supply of high-quality food. In evolutionary terms each acts in their own self-interest, just as a dairy farmer who takes good care of his cattle and supplies them with the food they need in order to produce plenty of high-quality milk is acting in his own self-interest. What makes the termite/fungus partnership so successful is that like all mature ecosystems it involves loops and feedbacks that retain the flow of nutrients within the loops.

Critics of Gaia theory respond to such examples by saying that this is 'only symbiosis'. Gaia theorists respond to that criticism by pointing out that that is the whole point – Gaia is symbiosis on a planetary scale. Or as Turner puts it, 'the collective pursuit of genetic self-interest among the partners in a symbiosis is no longer such a controversial idea.'

It is, though, a big step from a termite mound to a planetary symbiosis. But, fortunately for anyone wanting to demonstrate Gaian influences at work on a more impressive scale than a termite mound, there is an example of an ecosystem which operates on Gaian principles and is not only big in its own right, but is demonstrably connected to systems that affect the entire planet – the Amazon rainforest.

The British ecology writer Peter Bunyard presented the evidence for links between climate and the Amazon rainforest as part of a Gaian system at the Gaia 2000 meeting in Valencia. The most important link is that the rainforest makes clouds by releasing water vapour into the air through its leaves in evapotranspiration. This adds to the moisture in the air from other sources, in particular from the evaporation of water out at sea. But a great deal of the water that has seemingly been 'lost' to the forest in this way falls back from the clouds as rain – another example of the efficient recycling of resources in a mature ecosystem. At the most basic level, if the forest disappeared overnight, there would be such a great reduction in rainfall that the forest could never grow back. So, clearly, the link is important. But there are much more subtle effects going on, and their influence extends far beyond the Amazon Basin.

Because the prevailing winds in that part of the world blow from east to west, from the Atlantic to the Andes, most of the recycled water falls as rain inland, eventually making its way into rivers and down towards the Atlantic Ocean. Several studies show that about half the total rainfall over the basin eventually gets into the Amazon River and out into the Atlantic, while perhaps a quarter stays in the basin where it is recycled through evapotranspiration and local precipitation. That leaves another 25 per cent or so of the moisture which is carried out of the basin on the winds, some north towards the Caribbean, some south towards São Paulo, some even circling south and east towards South Africa, and a tiny proportion getting over the Andes to the Pacific coast of South America. The dominant role of the Amazon Basin in the region is such that although only a modest proportion of the water vapour from the Amazon Basin is carried south, it provides about 70 per cent of the average annual rainfall in São Paulo, the richest state in Brazil, and half the rainfall in Argentina. Bunyard points out that in the early years of this century both São Paulo and Rio de Janeiro have suffered repeated electrical blackouts as a result of falling water levels in their hydroelectric reservoirs. Like others, he suggests that these problems are being caused by deforestation in the Amazon Basin, which is already starting to influence the climate of the region.

In fact, the whole world is likely to be affected if this goes on. The amount of solar energy which is needed to drive the evapotranspiration process in the Amazon Basin has been estimated at about 520 Terawatts, which is some 40 times the total amount of energy used by humankind each year at the beginning of the twenty-first century. The Amazon Basin, says Bunyard, is 'a hydrological power engine' which is 'a critical component of contemporary climate'. And not just contemporary climate. It used to be thought that during an Ice Age, when the region was both cooler and drier, the Amazon forest would have been replaced by grassy savanna; but now pollen records found in deposits going back more than ten million years show that there was still widespread forest across the basin during Ice Ages. This discovery led to studies which show that although the forest would disappear if the region dried out but stayed as warm as it is today, when the temperature is lower it not only can adapt to reduced rainfall – partly because the trees can more easily retain moisture when it is cooler – but can also cope with the lower concentration of carbon dioxide in the air during an Ice Age. But the same studies show that if temperatures rise and rainfall decreases, the forest will be in serious trouble.

The energy pumped into the climate system by the Amazonian 'hydrological power engine' pushes air around the globe. The solar energy used to evaporate water is released again as heat when the water vapour condenses back into water, powering convection and creating things like thunderstorms, and the energy spreading out in this way from the Amazon Basin affects the climate not only in South America but in North America, South Africa, Southeast Asia and Europe, through a series of so-called 'teleconnections'. In addition, water vapour is a potent greenhouse gas, and although it is recycled through the atmosphere rapidly, changes in the average amount of water vapour in the air and the distribution of the vapour from place to place are bound to affect climate, although as yet our computer models are not sophisticated enough to predict just what the effects will be. Some of those models are good enough, though, to pinpoint some specific impacts of the loss of the Brazilian rainforest. For example, Roni Avissar, Head of the Department of Civil and Environmental Engineering at Duke University, has shown that there would be a significant reduction in rainfall – as much as 20 per cent – across the corn belt of the United States. There is no need to spell out the consequences of such a permanent drought in the region dubbed 'the breadbasket of the world'.

The models also suggest that the decline in rainfall may occur suddenly, following a period in which precipitation in the region around the rainforest actually goes up. As patches of rainforest are cleared and turned over to crops or grassland for pasture, they form dry 'islands' in the forest which absorb more energy from the Sun than their surroundings do, and get hotter than their surroundings. As a consequence, the air in the clearings rises through convection, drawing in moist air from their surrounding forest, which joins the rising column, carrying moisture away to form clouds and then rain.

Avissar suggests that this process will be at work for clearings up to about a hundred kilometres across, but that as the clearings become bigger and the forest becomes more fragmented at some point there will be a dramatic breakdown of the cloud-formation process. His best guess is that as well as each clearing being less than a hundred kilometres wide, overall the region has to be 60 per cent covered by forest for the pattern to be maintained.

Even without forest clearance, the Amazon rainforest is at risk today – from global warming. When the eastern Pacific Ocean warms, rainfall over the Amazon Basin declines – something that happens naturally in years when the weather pattern known as El Niño sets in. In the El Niño year 1998, the soil of the Brazilian rainforest dried out to a depth of five metres, and the trees essentially stopped growing. The forest has evolved to cope with occasional El Niño years, and soon recovered. But the models used by the UK's Hadley Centre suggest that as the world gets warmer, the El Niño pattern will become normal rather than exceptional. This will cause the forest to die back, reducing evapotranspiration and making the situation worse. This is likely to happen within fifty years, at the present rate of global warming. As the forest dies, huge quantities of carbon will be released from the vegetation and the soil in the form of carbon dioxide, adding to the buildup of greenhouse gases in the atmosphere.

The role of the Amazon rainforest provides us with both an example of Gaian processes at work, and a hint of what may happen to the world if human activities disrupt those processes. It is time to look at the way Gaia has coped with catastrophes in the past, and at what we might do to help Gaia cope with the catastrophe that threatens her, and us, now. Bearing in mind, of course, that what is good for Gaia may not be good for us.

10

Coping with Catastrophe

Near the end of *Homage to Gaia*, in a passage written shortly before his eightieth birthday, Lovelock contemplates a quiet retirement in the company of Sandy:

> We feel that the rest of our years together should be free of the tasks that we have disliked, but felt the need to do for duty's sake. Prominent among these for me is lecturing and attending meetings.

For half a decade, the Lovelocks enjoyed the quiet life. Their major achievement during that time, one of which both of them are hugely and justifiably proud, was walking the 630 miles of heritage path around the southwest coast of England, from Poole in Dorset along the south coast to Land's End and on along the northern coast to Minehead in Somerset. They didn't attempt it in one go, but walked the path in sections, staying at different cottages along the route from time to time and walking the local section of path, filling in the pieces over the years. Jim planned everything with meticulous scientific precision. The walk involves a total climb of nearly 33,000 metres, more than three times the height of Everest, and an equivalent descent; among other things, he checked to make sure they tackled the steepest ascents when the wind would be behind them. 'It was,' he says with a smile, 'a pointless but joyous personal challenge.' Sandy sees things from a slightly different perspective: 'I and others are amazed by the way he must deal with repeated physical ailments on an almost daily basis while managing a heavy workload that would defeat someone much younger and healthier than him. Giving up, or giving in to illness, is not what he is about.' The next phase of Lovelock's life proves her point.

In 2006, Lovelock suddenly reappeared centre stage in the debate about the future of humankind, with the publication of his most polemical book, *The Revenge of Gaia*. For the next two years, he returned to the self-imposed task of lecturing and attending meetings to hammer home the message that Gaia is in crisis as a result of global warming, and that the best we can hope for is an orderly retreat from our present destructive activities to ensure the survival of civilization at a greatly reduced level of population. The worst could see humankind reduced to 'a few breeding pairs'. He was so busy in 2006 and 2007, even at the age of 87, that in the months we spent working on this book we had to squeeze our meetings with him in between his visits to the United States, Japan, Norway, Spain and Australia, as well as a hectic schedule of engagements in the UK. What happened to bring about this change of heart?

'I first realized that we were on the brink of disaster in January 2004,' he says. During his 'retirement', Lovelock had become interested in the way the Intergovernmental Panel on Climate Change (IPCC), the world's most authoritative voice on such things, gathered its evidence and drew its conclusions. This interest was partly stimulated by the publication of a major report by the IPCC in 2001. With his network of contacts, and a name that opens many doors, Lovelock arranged a visit with Sandy to the UK's principal climate research centre, the Hadley Centre, which happens to be at Exeter, not far from the Lovelocks' home. There, they talked to various scientists about their work. Some were concerned by the melting of Arctic sea ice; others studied Greenland's vanishing glaciers; in another office, researchers investigated the effect of global warming on the tropical forests. At the end of the day, Jim and Sandy learned about the changes taking place in the great boreal forests of Siberia and Canada as the world warmed, and in return they talked of Jim's concern that life in the oceans would suffer as the temperature of surface waters increased.

Each item on its own provided evidence of positive feedbacks already acting to speed the rate of change. But 'there was no pressing sense of urgency. The researches were presented in a detached manner, as if the scientists concerned were discussing another planet. They didn't seem to understand emotionally that the changes they were describing were already affecting the lives of real people. What was worse, the scientists studying the Greenland glaciers, for example, only seemed aware in a general sense of the work on tropical forests. Nobody understood that the whole picture was far more serious than any of its components. And far more serious than anything I had read about even in what had previously seemed rather alarmist stories in the media. When I got home and re-read the IPCC report in the light of what I had learned, it scared me stiff. I talked it over with Sandy, and the upshot was that I wrote *Revenge*. All I really did was take the IPCC report and present their results in a way comprehensible to the non-scientist. I certainly intended it as a wake-up call.'

'I'm not a pessimist,' he continued. 'I'd spent forty years working with the concept of Gaia, and I felt comfortable with her. But now I realized just how gravely we had damaged the planet, and that there was a real threat that she would punish us with extinction. We have made an appalling mess, but the party that was the twentieth century is over and it is time to clear up that mess.'

Global warming is only part of the problem, although it is the most pressing part. Lovelock points to studies by the biologist Edward O. Wilson and others, which show that by destroying natural habitats human actions are now causing a loss of species on Earth comparable to the extinction of life that occurred at the time of the 'death of the dinosaurs', some 65 million years ago. That is a good place to start in trying to assess the coming catastrophe. Gaia has coped with catastrophes in the past, and may be able to cope with the catastrophe we are causing – but she may do so at our expense.

In fact, there have been five major extinction events on Earth since life emerged from the sea on to the land – five global catastrophes – and a host of lesser disasters. Death is a way of life on Earth, and the fossil record shows that even during the quiet times between extinction events, one species of life on Earth dies out every four years or so. This is the natural background level of extinction when all is well with Gaia. Looking at this another way, the average lifetime of a species is a few million years. But although species disappear from the record, they can leave descendants, because they have evolved into something else. Birds, for example, evolved from dinosaurs. The two biggest extinction events in the fossil record occurred at the end of the Permian Period of geological time,[1] about 250 million years ago, and at the end of the Cretaceous, about 65 million years ago. In the Permian extinction, about 90 per cent of all species living in the sea were wiped out; in the Cretaceous extinction (sometimes called the 'Terminal Cretaceous Event') about 70 per cent of marine species died, together with at least 50 per cent of species on land, including the dinosaurs.

The other three members of the 'Big Five' were only marginally less dramatic. At the end of the Devonian, some 360 million years ago, 70 per cent of all species disappeared; but this event is ranked slightly below the two worst disasters because the extinctions were spread over an interval of about 20 million years, in a series of lesser events. About 200 million years ago, at the end of the Triassic, about 20 per cent of all marine families (not just species!) disappeared, taking with them the last of the large amphibians and the therapsids, the predecessors of the dinosaurs. The first member of the Big Five chronologically occurred about 440 million years ago, at the end of the Ordovician, in the form of a pair of events which together may have killed off as many species as the Terminal Cretaceous Event.

Nobody can say for sure exactly why these events happened, although in every case there seems to have been a long-term pressure of some kind affecting life on Earth – perhaps climate change – followed by a sudden event, such as an asteroid impact or a huge series of volcanic eruptions, which brought catastrophe to life forms that were already under stress. The sudden events on their own do not seem to be sufficient to cause such damage to a completely healthy Earth system. We know, for example, that there have been very many asteroid and comet impacts on the Earth which have left scars on the planet's surface and caused local loss of life without disrupting things on a global scale. But when Gaia is already a little off colour, a sudden catastrophe can be the last straw that triggers a major extinction event. This is almost certainly what happened 65 million years ago. Because the Terminal Cretaceous Event was the most recent of the

1. It is no coincidence that major extinctions occurred at the end of geological time intervals; the geologists use these markers in the fossil record to define the different Periods and other subdivisions of time.

Big Five, it is the one for which we have the best geological and fossil evidence, so it is quite clear that dinosaurs and other species were already suffering some kind of stress when an asteroid about 10 km across struck the planet; we even know where it struck, since the buried crater has been found at Chicxulub on the Yucatan Peninsula of Mexico.

What could cause the long-term pressure? Obvious candidates include epochs of increased volcanic activity associated with the processes of continental drift and plate tectonics, widespread falls in sea level occurring when continents are raised up by geological processes, long-term Ice Ages or sustained global warming, and the formation through continental drift of supercontinents which dry out in their interior. But the causes of extinction events are not our concern here; rather, we wish to set them in the context of Gaia and the human impact on Gaia.

Some unthinking critics of Gaia theory have pointed to these extinctions as evidence that the theory must be wrong, since Gaia is unable to sustain perfect conditions for life throughout all of geological time. 'This does annoy me,' says Lovelock. 'It's like saying that because a person suffers a bout of 'flu, they aren't alive. The whole point is that the planet recovers from these events, even disasters in which 90 per cent of species have been wiped out. To me, this shows how effective Gaia is at bringing the Earth back from the brink of catastrophe to a healthy state. That's what homeostasis is all about. It means that a system returns to its optimum state even after it has been perturbed.'

So how does the 'perturbation' that we are applying to the Earth System – to Gaia – match up to these previous events? The figures are alarming, and ram home the point that it isn't just global warming that we need to be worrying about.

It's hard to be sure just how many species have been vanishing from our planet in the early years of the twenty-first century, but experts such as Richard Leakey and Edward O. Wilson have estimated that in round numbers it is 50,000 a year. That's 200,000 times faster than the background rate of extinction when Gaia is in good health. At one level, this huge number gives you some kind of feeling for just how many different species there are on Earth even today; at another, it is so vast that even if the estimate is ten, or even a hundred, times too big (and there is no evidence that it is) the rate at which species are being lost would still count as a major extinction. We are, indeed, living through a sixth extinction to rank with the Big Five – and this time, it is being caused by human activities, including global warming but also the destruction of natural habitat to make room for farmland and urban development. The *rate* at which species are being lost is already faster than during any of the Big Five extinctions,[1] and if it continues, 50 per cent of all species of life on Earth will become extinct by the end of the present century.[2] If that happens, we will have done as much damage to Gaia as the impact of the asteroid that killed the dinosaurs. 'The death of the dinosaurs,' Lovelock points out, 'opened the way for the rise of the mammals, including ourselves. At the rate we are going, the death of the mammals will soon be opening the way for the rise of some other varieties of life on Earth.' Even if the world were not getting warmer, he emphasizes, this loss of life would still be a disaster for us and – in the short term – for Gaia.

But in the context of global warming, the immediate threat facing humankind, the short-term importance of this sixth extinction is that it is reducing Gaia's ability to compensate for the rising temperatures. The probable result is that instead of settling back down into the kind of temperature regime that we think of as normal once the crisis has passed, Gaia is likely to switch into a different stable state, maintained by homeostasis, with temperatures several degrees higher than those of today. And the flip could happen in the space of a few years, within a few decades from now. This is the nightmare scenario that encouraged Lovelock to issue his wake-up call.

There is a simple example of how such a flip from one stable state to another can occur in the real world, and it is one which happens to be relevant to the present problem. The reason why a cap of floating ice covers the Arctic Ocean is partly that the North Pole lies almost in the middle of a shallow sea which is itself almost entirely surrounded by land. Warm currents from the tropics cannot penetrate

1. Even an 'instantaneous' disaster like an asteroid impact actually takes many decades to spread its influence throughout the worldwide ecological web.

2. Up-to-date information and links relating to the sixth extinction can be found at *www.well.com/user/davidu/extinction.html*

into the Arctic Ocean to melt the ice, which has stayed there since the Earth was in a much colder state during the latest Ice Age. But the other part of the story is that the ice is still there after all this time, even though the world has got warmer, because of the way it influences its own environment. Because the ice is shiny and reflects away a great deal of solar energy (it has a high albedo, more than 0.8), it persists even through the long Arctic summer when the Sun doesn't set. But if the ice cover were to go, the dark ocean would absorb enough energy each summer to warm to the point where the ice could not re-form, except possibly as a thin layer during the long Arctic winter when the sun never rises, even though the warm equatorial currents were still prevented from penetrating far to the north. The more heat that is stored in summer, the thinner the ice is in winter.

The loss of Arctic sea ice is already happening. The area of summer sea ice has shrunk by 8 per cent per decade since the 1970s, and thinned by about a metre between 1987 and 1997. Heat also escapes more easily from the open ocean to warm the air above it, which is one reason why the Arctic is warming rapidly today. This includes both a steady warming and occasional very extreme heatwaves – Lovelock points out that in April 2006 the Arctic archipelago of Svalbard was 12°C warmer than the long-term average.

The crucial point is that once the ice melts, you would have to cool the Arctic well below its present-day temperature for the ice sheet to re-form. For the present-day pattern of global temperatures, the North polar region can exist in a warm, stable state with no ice, or in a cool, stable state with ice. But it can't exist in an in-between state. Once the area covered by ice is reduced to some critical point, more heat will be absorbed in summer than is being lost in winter and the entire ice cap will rapidly disappear, at least in summer. This is an example of a 'tipping point', a moment when the internal dynamics of a system take over and continue a trend originally started by external forces. Conversely, in a cooling world there will be little or no ice cover until a tipping point is reached at temperatures much lower than those of today, when the ice will quickly spread to cover the entire polar sea.

The loss of sea ice that is already occurring has widespread implications. Researchers from the University of East Anglia suggest that the change in Northern Hemisphere temperature patterns will alter the path of the jet stream, and a team at the University of California, Santa Cruz, has shown that one result of this could be to shift storm tracks over North America, causing the western United States to dry out.

All this would not matter so much if the change continued in a steady fashion, even at a rate of 8 per cent per decade. But the heart of Lovelock's message in *Revenge* is that such changes will occur suddenly, and soon, as the various tipping points are reached. The example of shrinking Arctic ice highlights how a smooth change in some external forcing (in this case, temperature) does not necessarily produce a smooth change in some important property of Gaia such as ice cover – an example in miniature of what Lovelock believes is about to happen to the whole Earth. But it is more than just a 'thought experiment'. Within the context of global climate change, there is considerable evidence that the melting back of Arctic sea ice in summer is now at the tipping point, so that we may see an ice-free Arctic Ocean very soon.

Until recently, the best estimates of when this might happen could say only that the Arctic was likely to become ice-free in September – the key indicator – by the end of the twenty-first century. But at the end of 2006, researchers from the US National Center for Atmospheric Research, at the University of Washington, Seattle, and the Lamont Doherty Earth Observatory of Columbia University in New York, reported a series of computer simulations which take full account of the way ice loss is accelerating. As is the usual practice with such simulations, the model was run several times with slightly different starting conditions to see how reliable the simulations are. In the most extreme of seven such runs, the sea ice declined steadily until 2040, then collapsed from 6 million to 2 million square kilometres in a decade. The other runs were only slightly less dramatic, with the simulations taking only 5 to 10 years longer to become nearly ice-free in September. Cecilia Blitz, one of the members of the team, said in January 2007 that 'the rapid declines seen in our runs are a serious concern,' and suggest that the Arctic could be ice-free 20 to 40 years sooner than earlier studies, largely based on extrapolating the trend in ice cover over recent decades, had suggested. She also pointed out that their scenario assumes only modest increases in emissions of greenhouse gases; anything like 'business as usual' and the ice will melt even sooner. But if emissions are fixed at the levels of the year 2000, 'the sea ice retreats for only another decade or two and then levels off.' There seems little prospect of that. It is far more likely that the Arctic Ocean will flip into an ice-free state.

Working with Lee Kump, an Earth scientist at Penn State University, Lovelock has developed a model which shows how a similar flip involving the global ocean itself is likely to affect the whole planet in the middle of the twenty-first century. This model is in a sense a descendant of Daisyworld, but, with improved computers to do the modelling and a better understanding of how the components of Gaia interact, it is intended to demonstrate what happens in the real world, not on some hypothetical planet where daisies reign supreme.

The model describes a planet with land masses and oceans in the same proportions as on Earth, at the same distance from a star with the same luminosity as the Sun. It has algae growing in the sea, and plants growing on the land – as in the real world, life in the ocean is far more important to this model Gaia, in spite of our parochial impression that life on land is what matters. The temperature of this model planet depends on the amount of heat it receives from sunlight, and the amount of carbon dioxide in the air, which itself depends on the area of the planet covered by land plants and algae.

If they are given ideal growing conditions, in controlled environments both forms of life show increasing growth rates as temperatures rise from 0°C to 30°C, then a decline until they die out at 50°C. But there is a crucial difference in the real world. In the ocean, at a temperature of about 10°C to 12°C a layer of warm water several tens of metres thick becomes established on the surface of the sea, and this acts like a lid, preventing the upwelling which would otherwise bring to the surface the deeper water that contains the nutrients algae live on. In the surface layer, there is sunlight but no nutrients; in the deep water, there is nutrient but no sunlight. Algae need both to survive.

Something similar happens on land. Plant growth increases as the temperature rises until it reaches about 22°C. Above this temperature, more water is lost through evaporation than falls as rain, so the land dries out and the plants die. As we have seen, tropical rainforest is an exception to this rule because it has evolved mechanisms to recycle water, but conditions suitable for this to occur are indeed the exception rather than the rule.

Lovelock and Kump started their model running with a concentration of carbon dioxide in the atmosphere of 280 parts per million, corresponding to the conditions in the Earth's atmosphere before the Industrial Revolution, then gradually increased it to mimic the effect of anthropogenic emissions. The model is sophisticated enough to allow for seasonal variations and the difference in temperature at different latitudes, and shows a steady increase in temperature very much in line with what has been observed on Earth over the past century until the carbon dioxide concentration reaches 500 parts per million. At that point, the surface of the ocean becomes warm enough to form a lid and the algae die, causing a release of carbon dioxide which produces a jump in temperature of 6°C to a new stable state, where conditions are regulated by the surviving land plants. As with the example of Arctic sea ice, once the hot state is reached, reducing the external forcing – in this case, carbon dioxide – does not immediately flip the world back into the cooler state. It takes a very long time for the surface layer of the ocean to cool and for the algae to recover.

Lovelock is the first to point out that this model is a simplification, but the key role of marine organisms in regulating carbon dioxide, and the development of a layer of warm water over the oceans, are both factors which apply in the real world. 'The models of the IPCC report all suggest a steady rise in temperature as the carbon dioxide concentration increases over the next few decades. But our model predicts a sudden and rapid jump to a stable but hotter state, with no easy way back.' Simulations like those of the IPCC are based on the idea that the sensitivity of the system to a doubling of the amount of carbon dioxide in the air is the same wherever you start from – in the case of the IPCC models, a 3.2°C rise in temperature. This is largely based on observations of the real world, and is largely due to feedbacks, such as those involving water vapour. But Lovelock says this only applies to concentrations of carbon dioxide of a few hundred parts per million. He contends that once the concentration rises to about 1000 ppm, the link is broken; as the carbon dioxide abundance rises further the temperature no longer rises in proportion.

This prediction of the model is given added weight by a study of what seems to have been just such an abrupt jump in temperature, associated with a buildup of carbon dioxide in the air, that occurred on Earth some 55 million years ago. Lovelock cites the work of geologist Harry Elderfield, of Cambridge University, who has shown from an analysis of carbon and oxygen isotopes in sedimentary rocks from that time, early in the Eocene, that the Earth warmed by several degrees as the atmosphere was flooded with greenhouse gases in the form of carbon compounds. Altogether about two teratonnes (two million million tonnes) of carbon was put into the air in the form of these compounds; just why this happened is unclear, but a plausible explanation is that it was initially in the form of methane, a powerful greenhouse gas, released from seabed deposits known as clathrates. The greenhouse effect of the methane would have produced a rapid global warming; then, as the compounds oxidized to make carbon dioxide and water vapour, the carbon dioxide would have maintained a blanket around the Earth for a long time.

Elderfield finds that at the time the carbon compounds were injected into the air the temperature in the tropics rose by 5°C, while higher latitudes warmed by as much as 8°C. The whole process took place over an interval of about 10,000 years, compared with the couple of hundred years we are taking to do something similar to the atmosphere. But it then took 200,000 years for Gaia to recover, as the natural process of rock weathering we described earlier drew carbon dioxide down out of the atmosphere. According to Elderfield, this happened even though the concentration of carbon dioxide in the atmosphere during this Eocene event may never have risen above 440 parts per million, partly because methane, a stronger greenhouse gas than carbon dioxide, was involved. According to Lovelock, 'In thirty years, if we continue business as usual, we will have added a thousand teratonnes of carbon to the atmosphere as carbon dioxide and increased the carbon dioxide concentration to 600 parts per million. And we are adding other greenhouse gases, including methane, as well. The speed with which we are doing this gives the Earth System little time to adjust, and this causes other problems.'

One of those problems could be the very thing that caused the Eocene warming – methane. There are vast amounts of methane, produced by the decay of organic remains, stored in bogs around the world. It used to be known as marsh gas, and can sometimes ignite spontaneously to produce flickering will o' the wisp flames that terrified our superstitious ancestors. In Siberia alone the area covered by such bogs is as large as Germany, France and the Great Britain combined. Alarmingly, although much of this huge area, as well as similar bogs in Alaska, has been frozen, trapping the methane it contains, it is now thawing as the world warms. The bogs are drying out and releasing methane, which enhances the greenhouse effect and leads to more rapid warming. 'Nobody can really calculate how big this extra effect is likely to be,' says Lovelock, 'but we know that it will make a bad situation worse. And if the warming triggered a release of methane from clathrates, it would be worse still than the Eocene warming.'

Another problem is that carbon dioxide reacts with the surface water of the ocean to make carbonic acid, and if the acidity of the ocean becomes too great this will kill the organisms that usually take carbon dioxide out of the air and use it to build their carbonate shells. These include creatures known as coccolithophorids, tiny plankton at the bottom of the food chain. As well as having consequences for fisheries and for the buildup of carbon dioxide in the air, coccolithophorids are important contributors of dimethyl sulphide to the atmosphere. The complicated feedbacks have not yet been worked out in detail, but Lovelock draws attention to the work of Carol Turley, of the Plymouth Marine Laboratories, who has suggested that the ocean is already becoming too acid for the comfort of these marine organisms and that increasing acidification of the oceans will cause a decline in DMS and therefore a decrease in cloud cover, which will lead to increased global warming. In 2003, Ken Caldeira, of the Carnegie Institution in Stanford, and Michael Wickett, of the Lawrence Livermore Laboratory, estimated that over the next few centuries the oceans will become more acidic than they have been for 300 million years, except during extinction events such as the death of the dinosaurs.

As is so often the case, what matters here is not so much the size of the change being caused by human activities, but its speed. When the carbon dioxide built up more slowly 55 million years ago, there was time for the natural upwelling of more alkaline deep waters to neutralize this acid. But this ocean mixing takes thousands, not hundreds, of years. A Royal Society report published in 2005 spelled out how much faster the acidification we are seeing is than natural changes in the past, and concluded that the enhanced acidity will persist for thousands of years; the ocean acid spike expected in the *present* century will be more intense than anything the seas have experienced for at least 800,000 years. Once we reach the point where acid in the seas causes carbonate-forming creatures, including coral reefs, to die, more carbon dioxide will stay in the air and the greenhouse effect will get stronger faster. A team at the French Laboratory of Sciences of the Climate and Environment calculated in 2006 that a tipping point resulting in the dissolution of the shells of marine creatures known as pteropods, crucial ingredients of the food chain for whales, salmon and cod, will occur by 2050.

We discussed the acidity problem, and Gaia, with Richard Somerville, a Distinguished Professor at the Scripps Institution of Oceanography. He is among the researchers who regard Gaia as a powerful metaphor and valuable teaching aid, but as essentially a reworking of established ideas about the Earth. 'I emphatically share Lovelock's view that we Earth scientists are planetary physicians, and our patient has a fever, symptomatic of a serious disease.' He describes *The Revenge of Gaia* as 'excellent and worthwhile', and feels the same about *The Ages of Gaia*, even though he suggests that the message conveyed in both books 'does not require the packaging [of Gaia]'. He describes ocean acidification as 'one of the surest and most damaging consequences of adding carbon dioxide to the atmosphere', regardless of the threat of global warming.

But the warming now is also likely to be worse than the Eocene warming. There are already two important differences between the Earth today and the Earth as it was in the early Eocene. The Sun is about 0.5 per cent hotter than it was then, corresponding to an increase in global mean temperature of about half a degree, Celsius. And about half of the Earth's natural forest cover has been converted into farmland, or even into scrub and desert, by human activities, affecting Gaia's capacity for self-regulation. The analogy between the Earth System and a living cell highlights why this matters. As Ken Muldrew explained to us when we discussed Lovelock's work on cryobiology with him:

A cell can have myriad paths taking it between life and non-life when subjected to some harsh physico-chemical insult (such as a freeze–thaw cycle) but will only remain viable if a sufficient collection of interacting systems remain to enact the necessary repairs. *The geosphere, as well, is robust in the face of perturbations only when the complexity remains above some critical value* (with very little regard to the survival of any particular organism or isolated ecosystem). It is very much the stability and the complexity of the system that defines the living state rather than textbook particulars such as reproduction, growth and the like.[1]

'This is the very essence of Gaia theory,' says Lovelock. 'But don't imagine that we are talking about a complete breakdown of the Earth System. All the evidence is that Gaia will survive. She is a tough old bird. Even people will survive, in greatly reduced numbers. What people don't seem to get is the message that what is at risk is civilization.'

Lovelock's frustration at the way people fail to appreciate the imminence of the threat boiled over in an email we received from him in March 2007:

> Have you seen this week's *Nature*? There is an account of the rapid melting of Arctic floating ice and the author says that in twenty years the Arctic Ocean will be ice free in the summer. This is truly apocalyptic, yet still they see it as a regional not global problem.

Part of the reason for Lovelock's concern, once again, is feedback. As we have mentioned, as the Arctic warms, the permafrost at high latitudes thaws. These frozen Arctic soils are packed with organic matter – dead leaves, roots and animal remains – which has been preserved like food in a freezer. As it thaws, the organic matter decays, and both carbon dioxide and methane are released into the air, enhancing the greenhouse effect and increasing the rate of global warming and Arctic thawing. But this isn't all. There are also oil and gas deposits underneath the Arctic. A report in the *Independent* on 6 November 2006 noted that fifteen oil companies had applied to the US Department of the Interior for permission to explore the

1. Our emphasis.

US-controlled regions of the thawing Arctic. When we pointed this article out to Lovelock, for once he was lost for words.

But he had plenty to say about the way extreme weather events are still widely perceived as temporary, not long-term, difficulties. In 2003, for example, Europe experienced a heat wave in which an estimated 30,000 people died. By taking meteorological records at face value, it is possible to infer that the odds against such a hot spell are 300,000 to 1, so that it is sometimes referred as being a 'once in 300 thousand years event'. But that doesn't mean that we can expect to wait 300,000 years before there is a similar heatwave. If a horse wins a race at 500 to 1, its odds shorten for the next race it runs. The occurrence of a 'once in 300,000 years summer', set against a background of rising average temperatures and other climate changes, means that events which *used* to be rare are becoming more common. 'Few people seem to realize that the IPCC models predict that by 2040 most summers in Europe will be as hot as the summer of 2003. This is just a straightforward extrapolation of the trend we can already see. Apart from the direct effects on human health, what will be the effect on crops and livestock? What will there be to eat?'

He also drew our attention to a paper published in *Science* in May 2007, in which a team of highly regarded climatologists made a comparison between the projections made in the IPCC scenarios and what has been happening to the real world since those projections were made. For every parameter – carbon dioxide concentration, temperature, and sea-level rise – the changes that are occurring in the real world are at or above the *highest* of the IPCC projections. The IPCC warnings are too conservative.

Lovelock believes that 'the Earth System is now in positive feedback and is moving ineluctably towards the stable hot state of past climates like the early Eocene. There is still a haze of pollution over the Northern Hemisphere that is keeping temperatures two or three degrees lower than they would be if it were not for global dimming. If we carry on burning fossil fuel, the greenhouse effect will intensify. If we stopped using fossil fuel tomorrow, the atmosphere would clear and the world would get warmer. We are damned if we do and damned if we don't. The only hope of any kind of smooth transition is to reduce emissions while lessening our destruction of natural forests; but even then we have to adapt to change, and we need to start adapting right now.'

Lovelock also worries about a feedback involving global dimming and rising temperatures. When we visited him in June 2007, he had just identified a possible reason why the summer of 2003 was so extremely warm in Europe. The warmth of an unusually hot year was exacerbated, he believes, by a reduction in the haze over the continent linked to the warming itself. His scenario starts with the observation that on a hazy day the haze clears as the Sun rises and the temperature goes up. This is because the relative humidity of the atmosphere declines, which means that although the amount of water in the air stays the same the size of the droplets making up the haze gets smaller. 'Exactly the same thing used to happen in pea soup smogs when I was a boy,' he recalled. 'The fog would be so thick outside that you couldn't see a your own feet, but inside the house the air was clear. It wasn't because the polluted air couldn't get in, but because the room was warmer so the polluting particles were smaller.' Could the same thing have happened on a continent-wide scale in 2003? And could something similar happen on a global scale by 2040? If so, we would then feel the full force of the anthropogenic greenhouse effect, with no protection from global dimming.

Another reason why Lovelock is so concerned about global dimming is that recent studies have shown that there is more to this process than pollution particles simply blocking heat from the Sun and radiating some of it back out into space. These tiny aerosol particles also act as seeds for the growth of the tiny droplets of water that make clouds. Without the pollution, there are fewer cloud seeds, and large drops of water grow around the seeds until they are big enough to fall as rain. With more seeds, the same amount of water is spread over many more, smaller droplets which do not fall to the ground but increase the reflectivity of the clouds. You can see this process at work on a clear day anywhere where the sky is criss-crossed by aircraft contrails. Because this increases the albedo of the Earth, so-called 'global dimming' actually makes the Earth brighter in photographs taken from space. And because the global dimming effect is uneven, with most of it affecting the Northern Hemisphere, one consequence in the second half of the twentieth century was a shift in rainfall between the hemispheres, explaining the prolonged droughts in the Sahel region of Africa in the 1970s and 1980s. It is no coincidence, says Lovelock, that as the pollution began to clear the rains returned to the Sahel, although not yet in their previous strength.

The warmer world Lovelock envisages, within the lifetimes of most people on Earth today, would not be a bad place to live. High latitudes that are now too cold to be comfortable will thaw, and what remains of the British Isles after the inevitable sea-level rise could be a pleasantly warm archipelago. The planet could perhaps support a human population of as many as a billion people living in a sustainable fashion. The problem is getting there from here. The present population of the planet is approaching seven billion, and many centres of population are close to sea level. It's hardly surprising, especially given the horrors that his generation has experienced, that Lovelock likens the situation to the crisis that faces a nation at time of all-out war. In his view the kind of global effort needed to cope with the present crisis is comparable to the national effort made by Britain in the darkest days of the Second World War, when the survival of the civilization that the nation represented took precedence over personal comfort and even over personal survival. He has practical suggestions as to how we might manage 'a sustainable retreat' from the position we are now in, and extending the military analogy points out that one test of a good general is knowing when it is time to retreat. Napoleon just escaped from Moscow, although he left it late; Hitler, at Stalingrad and elsewhere, never accepted the need for retreat.

The analogy with Napoleon at the gates of Moscow is part of Lovelock's 'wake-up call', and it's a sign of how desperate he sees the situation to be that he is willing to encourage technological fixes to alleviate the problem of global warming in the short term. These are not cures, he emphasizes, any more than dialysis is a cure for kidney failure. Nevertheless, 'who would refuse dialysis if it gave them a chance to live long enough for a donor kidney to become available?' But we shouldn't run away with the idea that human ingenuity can take over the control of the systems that compose Gaia and keep things in balance in the long term. The Earth System is simply too complicated for us to run. 'Look at what happened to Biosphere II,' says Lovelock. This was a supposedly self-contained artificial habitat built in the Arizona desert, where volunteers attempted to live in balance with the plants and animals sealed into the biosphere with them – simulating, to some extent, what conditions might be like in a large space station or a human colony on the Moon. 'The results were very instructive. They were never able to maintain the oxygen concentration at the right level without cheating by giving the people inside some extra to breathe, ants and cockroaches overran the place, and the only birds that did well were local ones that got in while the place was being built. All the birds they deliberately introduced died. The people inside would have died as well if the experiment hadn't been abandoned. If we can't manage a little habitat in Arizona, what chance have we of managing the whole Earth System?'

The technological fixes Lovelock endorses are not, therefore, really 'fixes' at all, but short-term measures, equivalent to dialysis, to enable humankind to carry out 'a sustainable retreat' from its present untenable position outside the metaphorical gates of Moscow. They fall into two categories – reducing the amount of heat arriving at the surface of the Earth from the Sun, and removing carbon dioxide (and other greenhouse gases) from the air, or from the gases from burning fossil fuels before they even get into the atmosphere at large.

One of the best ways to reduce the amount of solar heat reaching the ground or the sea is to increase the cloud cover of the Earth, so that more of the incoming energy is reflected back into space. One suggestion – which requires more work before being put into practice – is to create an aerosol of cloud condensation nuclei from sea water itself, causing the growth of mist or low cloud over the surface of the sea. A more practical and immediate possibility is to 'imitate the well-known cooling effect of volcanoes' by spreading an aerosol of tiny sulphuric acid droplets through the stratosphere. We know this will work, because it was one of the main causes of the global dimming that shielded the Northern Hemisphere from the full force of the anthropogenic greenhouse effect in the middle decades of the twentieth century – the 'human volcano'. The best way to increase global dimming, says Lovelock, would be to add sulphur compounds to aircraft fuel, at a concentration of between 0.1 per cent and 1 per cent – and if we couldn't quite start doing it tomorrow, we could start doing it in a matter of weeks. 'Let's face it, people aren't going to stop flying. So we might as well get some benefit from burning all that fossil fuel.' At present, sulphur compounds are *removed* from aviation fuel in order to reduce pollution around airports. 'We might have to put up with pollution at ground level to avoid something worse and buy us time to retreat.' We, or rather Gaia, could also cope with the resulting increased acidity of the oceans. Although it would not be a good thing, sulphuric acid does not have the same effect on living things as carbonic acid, so could be tolerated in the short term.

A longer-term and more dramatic, but technologically and economically feasible, proposal is to put a disc-shaped 'sunshield' in space, in an orbit between the Earth and the Sun, to block out a small percentage of the Sun's light. During a solar eclipse, the Moon can block out all the Sun's light even though it is much smaller than the Sun, because the Sun is much further away (in round numbers, the Moon is roughly one-four-hundredth the size of the Sun, but the Sun is four hundred times further away). In the same way, the sunshield would not have to be very big. Researchers from the Lawrence Livermore Laboratory in California have calculated that a disc about 11 kilometres across placed at a point where the gravitational pulls of the Sun and Earth cancel out (one of the so-called Lagrangian points) would do the trick. The disc would weigh about a hundred tonnes, but would be invisible from Earth, lost in the glare of the Sun.

There are other ways to block out heat from the Sun or to increase the albedo of the Earth, but Lovelock, quoting his colleague Peter Liss, of the University of East Anglia, points out the big snag with all of them. They do nothing to reduce the buildup of carbon dioxide in the air – if anything they might encourage complacency – and 'this means the carbonic acid concentration of the oceans would continue to increase, with all that implies.' And 'the longer we go on releasing carbon dioxide, the more effort we would have to make to block out the heat from the Sun. This can only be a short-term fix.' Which makes it important to consider ways to remove carbon dioxide from the air or from combustion products and hide it somewhere.

Assuming the gas can be removed from the air and liquefied (which it can be), one of the first suggestions for dealing with it was to pump it into the sea, where the liquid would sink down to the bottom and take thousands of years to re-emerge. But 'the acidity problem puts paid to that idea'. Carbon dioxide can also be pumped into exhausted oil and gas fields where it would be stored on geological timescales. But the scale of the effort required is immense. Lovelock is fond of pointing out that at 2005 levels of emission, the yearly output of carbon dioxide, if solidified, would create a mountain more than one and a half kilometres high and some three kilometres in radius. One of the best ways to make use of so much material, he suggests, is to follow up the proposal of the America Klaus Lackner, who says that carbon dioxide from the air could be combined with so-called serpentine rocks to make magnesium carbonate, a solid that can be used as a building material.

Environment writer Fred Pearce, who was one of the earliest to cover the Gaia 'story' and has followed it for three decades, comments that with these and other ideas there is 'a huge potential to end our reliance on carbon fuels. There are tipping points in the human system too. Technically, we can fix it; economically it is no big deal. Politically – that's the rub.'

Getting any of these processes for sequestering carbon dioxide in a stable form will require time and money, even if, as Pearce suggests, it is the kind of money the global economy can afford. One of the things that infuriates Lovelock about the official attitude to the problem is that in 2005 the UK Nuclear Decommissioning Authority said that over the next twenty-five years the country would have to spend about £60 billion decommissioning nuclear reactors. 'Madness. I can't believe they would spend so much on such a trivial risk. If they've got so much money to play with, they ought to be spending it on ways to "decommission" carbon dioxide.' This brings us to the aspect of *The Revenge of Gaia* that caused the biggest stir, although in fact it said nothing that Lovelock had not said in his earlier books, going right back to the 1970s – his dismissal of the popular (mis)conceptions about the hazards of nuclear radiation and his enthusiastic espousal of nuclear fission as the best short-term way to keep electricity flowing while reducing our consumption of fossil fuels.[1] 'My track record

1. 'Short term' is an important caveat. There is a limited amount of fissionable material available, and a long-term solution, such as fusion power, needs to be developed before the middle of the present century.

shows that I have always seen nuclear energy as a force more for good than for harm.'

Electricity is the lifeblood of modern civilization. The existing, tried and tested, nuclear technology, Lovelock says, can produce large amounts of electricity cleanly and safely at a competitive price – witness the fact that 80 per cent of France's electricity is generated by nuclear power stations and has been for decades. On a visit to the Le Havre nuclear processing site as guests of the French in 2007, Jim and Sandy were shown the room where all the long-term waste from the past forty years is stored, 'It's about as big as a small cinema, with the waste buried three metres underground and holes in the floor to lower the stuff down. The radiation level in the room was less than in our sitting room at Coombe Mill. And even when we stood alongside the water-filled pits where the hot stuff from all over the country is cooling off before being processed, the level was one microsievert per hour. It's trivial. If you fly on a commercial airliner you get a dose of five times that, seven if you go on a polar route. And I love telling my Green friends that when they travel on a French train it is using electricity from nuclear energy!'

It's worth quoting something Lovelock said in his first book, *Gaia*, published in 1979:

> It has been predicted that the increase in carbon dioxide will act as a sort of gaseous blanket to keep the Earth warmer. It has also been argued that the increase in haziness of the atmosphere might produce some cooling effect. It has even been suggested that at present these two effects cancel one another out and that is why nothing significant has so far emerged from the perturbation caused by the burning of fossil fuels. If the growth predictions are correct and if as time goes by our consumption of these fuels continues to double more or less with each passing decade, we shall need to be vigilant.

In the same book, he pooh-poohed the idea that nuclear radiation poses a serious threat to life on Earth. 'If all the nuclear weapons stock-piled on Earth were exploded almost simultaneously,' there would be detrimental effects on 'we and the larger animals and plants . . . but it is doubtful whether unicellular life would for the most part even notice.' On the Bikini Atoll where the United States carried out many nuclear explosions, there has been 'little effect on the normal ecology of the area, except in places where the explosions had blown away the top soil and left bare rock behind.'

Of course, most people are rightly worried about the fate of 'we and the larger animals and plants', not just the fate of unicellular life and Gaia. But here too the hazards of nuclear radiation have, Lovelock emphasizes, been wildly exaggerated, as we shall see shortly.

There are two ways to obtain energy from nuclear reactions. The nuclei of both light elements and heavy elements store more energy per particle than the nuclei of middling elements. So if very light nuclei can be persuaded to fuse together to make heavier elements, energy is released; similarly, if very heavy nuclei can be persuaded to split into lighter nuclei energy is released. The first process, nuclear fusion, is the process that keeps the Sun and other stars shining. Specifically, inside the Sun nuclei of the lightest element, hydrogen, fuse to form nuclei of helium, with energy being released. If this process could be reproduced on Earth, it would provide a source of clean, safe energy with no radioactive waste products (essentially, no waste products at all). Since hydrogen can be obtained in vast quantities from sea water, nuclear fusion is seen as the long-term solution to the world's energy problems. The operative word here, though, is 'long-term'. Even optimists see a working fusion power station as being at least thirty years away.[1]

The problem is that hydrogen fusion occurs in the heart of the Sun because it is so hot there (about 15 million°C) and the pressure is so immense (many times the density of lead). Because we cannot match the density at the heart of the Sun in reactors built on Earth, the temperature inside those reactors has to be even greater than that at the heart of the Sun to make fusion occur. The material in a fusion reactor, a form of hot gas called a plasma, has to be contained in a hot fireball, using intense magnetic fields, for long enough for fusion to begin, and all of this requires a large input of energy. The most successful hydrogen fusion reactor to date, called the Joint European Torus, or JET, has managed to obtain a power output of 16 mega-

1. The optimists have been seeing a working nuclear fusion power station as being about thirty years in the future ever since we started reading such forecasts in the 1950s.

watts for about a second – but this was only about two thirds of the energy put into the reactor to heat the contained plasma. But it did generate heat from nuclear fusion. The success of JET has led to an international collaboration to build an improved reactor, called ITER, in southeast France. With a budget of €10 billion, ITER is planned to start operating in 2016, providing 500 MW of power (more than it needs to 'light the flame'), and to run for 20 years as a test bed. The name, incidentally, comes from the acronym for 'International Thermonuclear Experimental Reactor'. But such is the public image associated with the words 'thermonuclear' and 'experimental' that it is now simply known as Iter, and in PR handouts we are told that this name comes from the Latin word for 'path', since Iter represents the path to nuclear fusion power.

Even if the utopian dream of a global civilization powered by nuclear fusion is to come true, there is a gap of several decades to fill with whatever sources of power we can find that do not release greenhouse gases. Although he accepts that every little helps, Lovelock is particularly scathing about 'the woolly-headed thinking' of some members of the Green movement who imagine that this could be done using so-called renewable energy sources. He points out that the use of wind energy, for example, is still at an early stage of development, that each 100 metre tall wind turbine tower generates only about a megawatt of power, and then only when the wind is blowing. 'To supply the energy needs of the UK today would require more than a quarter of a million windmills like this, packed in at a density of three every square mile if we exclude national parks, urban areas and industrial sites.' In *Revenge*, he quotes a Danish study which found that in practical terms wind power can never provide more than 3 per cent of the energy needs of modern European society. The most damning evidence comes from a report by the German Audit Commission, which examined the record of power generation by some 20,000 turbines across the country. It found that they operated effectively for only 18 per cent of the time, so that for 82 per cent of the time their share of the load had to be taken up by power stations burning fossil fuel, which were not even running at maximum efficiency because they had to adjust to the changing contribution from wind power. For an island like Britain, wave and tidal power are much better bets, and a barrage across the Severn estuary could provide 6 per cent of the United Kingdom's energy requirements. But all of this is just tinkering around the edges of the problem.

'The craziest idea is the notion that we can use biofuels as replacements for fossil fuels. All those people who run their cars on biofuel and think they are helping the environment are only deluding themselves. The fuel is made from plants grown on land that would otherwise be used to grow food, or left as forest. We have already taken more than half of the good land to grow food. It's not usually understood that farmland does not act like natural forest, and farming is almost as bad for Gaia as combustion. If we take more of the natural forest, Gaia will be badly affected. And if we turn all our agricultural land over to biofuel production, we will have nothing to eat.' Backing up these claims, he pointed to an article in the *Daily Mail* of 26 March 2007 which described how the last remaining tropical forest in Borneo, the home of the orang-utan, is at risk from the development of plantations of the African oil-palm, *Elais guineensis*, to produce oil that can be used as an 'environmentally friendly' fuel source. The destruction of the forest may be useless, because 'biodiesel derived from palm oil may actually produce twice as much CO_2 as Saudi crude, once the carbon dioxide released by forest clearance has been factored into the equation.' And he shakes his head on reading about Colorado Governor Bill Ritter's hopes of an 'energy economy' that will bring prosperity to his state from the production of ethanol. 'The process needs huge quantities of water, and Colorado is one of the regions getting drier as a result of global warming. The South Platte river is in serious decline.'

Lovelock is more approving of the use of solar energy to generate electricity, since in principle it would provide a clean source that did not require the use of fertile land. But the technology is still in its infancy, and as he put it in *Revenge*, 'I find it hard to believe that large-scale solar energy plants in desert regions, where the intensity and constancy of sunlight could be relied on, would compare in cost and reliability with fission or fusion energy, especially when the cost of transmitting the energy was taken into account.' Which brings us back to nuclear fission as the proven, reliable and safe method of generating large amounts of electricity in the places where it is required, as quickly as the power stations can be built.

Nuclear fission is easier to achieve than nuclear fusion because some heavy nuclei are naturally unstable and inclined to split into fragments, the nuclei of lighter elements, releasing energy as they do so. This occurs spontaneously for certain varieties (known as isotopes) of elements such as uranium and plutonium; the process can also be triggered by bombarding such unstable nuclei with high-energy particles. Since such high-energy particles are released when a nucleus splits, if enough radioactive uranium, for example, is gathered in one place, the particles released when one nucleus splits will trigger the splitting of one or more neighbouring nuclei, and so on in a self-sustaining chain reaction. In a fission bomb, each nucleus that splits triggers the splitting of several other nuclei in a runaway explosion. In a fission power reactor, material known as a moderator is used to slow down and stop many of the high-energy particles, so that on average each nucleus that splits triggers the splitting of just one other nucleus in a steady process that generates heat which can be used to make steam to drive turbines to produce electricity. So what's the problem?

There are two reasons why people worry about using nuclear fission to generate electricity in this way. The first is that the waste products from the process are themselves radioactive and emit energetic particles – nuclear radiation. The second is the fear that something could go badly wrong with the moderator, allowing a runaway nuclear explosion to occur. The second possibility is extraordinarily unlikely. In a fission bomb, the material that makes the nuclear explosion has to be squeezed together in a very dense compact lump, usually using conventional explosives, in order to trigger the runaway chain reaction. The worst that can happen with a nuclear power reactor is that it will get too hot and melt, as happened at Chernobyl in 1986. The molten material from the core of the reactor will flow and spread out, thinning the radioactive material and stopping the chain reaction. Chernobyl is, of course, the archetypal example used by opponents of the use of nuclear fission reactors to highlight what can go wrong and to demonstrate the hazards of radioactivity. But as a worst-case scenario, Lovelock finds it distinctly unimpressive. He turns their argument on its head, pointing out that although the region around Chernobyl has been abandoned by people, it is a flourishing wildlife habitat. Gaia doesn't seem at all bothered by the supposedly lethal radiation. Why?

It all goes back, Lovelock argues, to the early days of life on Earth. The radioactive material in the crust of the Earth today was produced inside one or more stars that exploded at the end of their lives and laced the interstellar clouds from which new stars and planets form with heavy elements.[1] These isotopes naturally decay into stable isotopes as they emit radiation, so the amount present in the crust declines as time passes. Today, the proportion of radioactive uranium-235 present in all the uranium on Earth is 0.72 per cent. But when the Earth formed, four and a half billion years ago, the proportion was 15 per cent, and other radioactive material was also present in richer concentrations then than it is now. In *Gaia*, published in 1979, Lovelock pointed out that 'life probably began under conditions of radioactivity far more intense than those which trouble the minds of certain present-day environmentalists.' In other words, life evolved to cope with much more background radiation than there is today. *Too much* radiation is lethal – drinking too much water is lethal – but the point is that the level at which danger occurs is much higher than most people think.

Lovelock returned to the theme in his most important book, *Ages of Gaia*, published in 1988. He credits a Dr Thomas, of Oak Ridge Associated Universities, with pointing out to him that oxygen is a potent cancer-causing agent, far more harmful to human beings than even the kind of radiation level that has driven people away from a wide area around Chernobyl. As far as cancer is concerned, 'breathing,' said Lovelock, 'is fifty times more dangerous than the sum total of radiation we normally receive from all sources.' It's the price we pay for the high-energy, active lifestyle that oxygen makes possible. In 2006, he spelled out why this is so in *The Revenge of Gaia*.

Lovelock's background in medical research means that he has a thorough understanding of the workings of the cell, and the way in

1. See our book *Stardust*, Allen Lane, 2000.

which cells use oxygen in reactions with food compounds – a form of slow burning – to release the energy which powers our bodies. Anywhere that oxygen is involved in such reactions, unwelcome by-products are released. These include a form of oxygen with an electron attached, called the superoxide ion, and the hydroxyl radical, which is a water molecule with one of its hydrogen atoms removed. These pollutants are highly reactive and can damage the key molecules of life, including DNA. The cell has evolved a suite of techniques to repair this damage, but no such process is perfect and in the long run in a few cells these repair mechanisms will fail and the failure will result in a cancer cell developing and spreading by reproducing itself. Lesser damage contributes to the general ageing process. This is the science behind the fad for taking so-called 'anti-oxidants' in the hope of slowing the ageing process; whether or not those commercial products do any good is another matter. What does matter is that, according to Lovelock, before reaching the age of seventy 30 per cent of all people die from some form of cancer, and the main cause of those cancers is breathing oxygen. Wild animals, of course, usually die from other causes long before they are old enough for this to be a problem.

This, argues Lovelock, is the background against which the threat from nuclear radiation should be assessed. In the extreme example – the worst we have – of survivors of the fission bomb dropped on Hiroshima, sixty years after being exposed to this intense blast of radiation the increased incidence of cancer among the survivors was 7 per cent. Not 7 per cent of the total number of survivors, but 7 per cent more than the proportion of cancer victims in the rest of the Japanese population, who had not been exposed to radiation. Nobody suggests that this is a good thing; but it is a much smaller risk than most people realize. The Chernobyl accident was much less damaging than the explosion of a fission bomb. Lovelock quotes figures from the World Health Organization showing that by 2005 just 75 people had died as a result, mostly the firemen and other emergency workers who tackled the fire in the reactor and were exposed to high doses of radiation. And yet a figure of 30,000 deaths resulting from the accident has often been quoted in the media. Why the discrepancy?

Lovelock points out that the figure comes from extrapolating the risk from the tiny exposure most people in the affected area received over their entire lifetimes. Everybody has to die some time, and in simple language the claim is that 30,000 people might die of cancer as a result of the Chernobyl accident rather than from, say, a heart attack. But *when*? Using data from the United Nations Scientific Committee on the Effects of Atomic Radiation, Lovelock calculates that the average effect of the radiation from Chernobyl on everyone exposed except the emergency workers is to reduce their life expectancy by between one and three *hours*. For comparison, he says, 'a life-long smoker will lose seven years of life'.

As with the smokers, some will die young and some will live to a ripe old age. But it is part of human nature to accept such risks. Thousands of people die on the roads every year, but we do not ban cars; even though flying is reputedly the safest way to travel, on occasion hundreds die in a single crash, but we do not stop flying. The death of 75 people as a result of one nuclear accident is no more reason to stop using nuclear power than the fact that an estimated 4,000 people died in floods from accidents in the hydro-electric industry between 1970 and 1992 is a reason to stop building dams.[1] Averaging over a population in which 30 per cent of people smoke cigarettes, smoking is 10,000 times more damaging than the Chernobyl accident. We can expect half a billion early deaths from smoking over the same time-frame as the one the Chernobyl scare stories are based on, and 60 million deaths worldwide in traffic accidents.

This also puts the vast sums being allocated to decommissioning nuclear power stations in another perspective. Lovelock has famously offered to take the radioactive waste material produced by a nuclear power station in a year and bury it on his land, where the heat from its continuing radioactive decay would help him to warm his home. He also suggests that if people are so scared of radioactivity, the waste could be scattered in the tropical forest, to frighten off developers. But the best proposal has come from scientists working at the European particle research centre, CERN, near Geneva. They have realized that the technology they use to accelerate beams of particles for research purposes could be used to 'burn' the radioactive waste from conventional reactors, and even the much more dangerous plutonium

1. Figures quoted by Lovelock from the Paul Scherrer Institute in Switzerland.

from dismantled nuclear weapons, to release more energy. In essence, particle beams could be directed at any radioactive substance, speeding up the rate of decay and releasing more energy than is needed to power the beams in the first place. It may sound far-fetched, but never underestimate the CERN scientists – among other things, they gave us the World Wide Web. Their proposed technique would be a clean, efficient way to dispose of nuclear waste of all kinds, hastening its conversion into stable, non-radioactive isotopes, with the added advantage that if anything went wrong and the beam stopped working everything would simply shut down. 'Radiation from nuclear waste can be dangerous,' says Lovelock, 'but only if you are foolish enough to swallow it or sit on top of it for a long time.'

It may already be too late to avert the looming climate catastrophe that Lovelock warns about, but he makes a powerful case that the only hope of averting the catastrophe is to cut back our use of energy and make as rapid a switch as possible from reliance on fossil fuels to nuclear fission. Without going quite as far as Lovelock in his support for nuclear power, Fred Pearce, who used to be opposed to the idea, now confesses 'I have become a revisionist. It seems to me we cannot ignore the nuclear option.' This echoes the view of many of the people we spoke to when researching this book, although some were reluctant to have their revisionism made public.

The 'Greens' are more in tune with Lovelock's other big idea – that we also need to get away from our obsession with economic growth, and understand when enough is enough. It's a tall order. Gaia will survive – she has coped with worse before. Human beings will survive, even in Lovelock's worst-case scenario. But what is at risk is human civilization. Sir Crispin Tickell hails the role of Lovelock's book in 'showing that self-regulation on the Gaian formula can sometimes be breached', and promoting awareness of what he prefers to refer to as 'climate destabilization'. Lovelock's erstwhile colleague Lynn Margulis agrees that *Revenge* is 'an important book', and that 'there is no way out of this without nuclear power. People should listen to Lovelock; his time has come.'

But you might expect people like Tickell and Margulis to say that. How can we put Lovelock's warnings in a broader perspective? How seriously should they be taken? One assessment was provided by a panel of seven experts convened by the BBC in response to the publication of *The Revenge of Gaia* in 2006.[1] The panel consisted of Brian Hoskins, of Reading University, Susan Owens, of Cambridge University, Ron Oxburgh, a former Chairman of Shell, Vicky Pope, of the Hadley Centre, Chris Rapley, head of the British Antarctic Survey, Hans von Storch, head of the German coastal research institute, and Andrew Watson, of the University of East Anglia. They agreed unanimously that:

- Global mean temperatures will rise by between 3°C and 5°C by 2100 'unless we act swiftly to cut greenhouse gas emissions and protect natural forest'
- Temperatures might rise by as much as 8°C by 2100
- A temperature rise of 3°C to 5°C 'would probably bring severe changes for humans'
- We are being reckless with the planet
- Population growth is a major issue
- The book is helpful in the climate debate
- Climate change is real, dangerous and significant in our own lifetimes.

Although six of the seven panellists agreed that 'James Lovelock is a towering figure in environment science and has been a major influence on understanding the way in which the Earth system works,' they disagreed with his claim that nuclear fission is (in the BBC's words) 'our only realistic short-term solution' and agreed that he 'is wrong in the book to reject wind power'.

This is probably a fair assessment of the response to Lovelock's book. The message from many quarters is that things are bad, but not quite as bad as Lovelock suggests. His response is that he would love to be proved wrong, but fears that such views are based on the discredited idea that changes will occur steadily as the temperature increases. 'It's frustrating that otherwise quite sensible people don't seem to appreciate the significance of the way the Earth System can suddenly jump into a hotter state when a critical point is reached.' Time will tell.

1. See *http://news.bbc.co.uk/1/hi/sci/tech/5152590.stm*

Andrew Watson's views are particularly relevant, because during his long association with Gaia theory he has not always seen eye to eye with Lovelock, and has never been afraid to speak his mind when he disagreed with his one-time PhD supervisor. 'We are now committed to substantial climate change, probably about as great as that between the last glacial maximum and the present day, but occurring a hundred times faster. Given the amount of positive feedback that we know is in the climate system anyway, and the possibility of unpleasant and unforecastable sudden changes of the kind we see in Daisyworld models, this is a scary situation to be in. We should be trying with all means available and as a matter of urgency to slow the forcing we are putting on the system.' In this context, 'The book acts as a counterbalance to those vested interests who have been arguing that there is absolutely nothing to worry about. The climate science in *Revenge of Gaia* is not outside the range of what, for example, IPCC Working Group I is forecasting, but the social consequences Jim foresees are more extreme. I hope he's wrong, but I'm not absolutely certain that he is.'

'I hate to be regarded as a pessimist, though,' Lovelock responds. 'I've always regarded myself as being something of an optimist.' With that in mind, in 2007 he came up with his own idea for a technological fix for global warming which goes far beyond the 'kidney dialysis' approach. His approach started from the idea of using the energy in the system to reverse one of the feedbacks contributing to global warming. He points out that the warm layer of water at the surface of the ocean is more salty than the layers beneath, because of evaporation, and that if it were cooled only slightly it would sink, because it would then be more dense than the water beneath. This happens naturally in the North Atlantic, as part of the process driving the circulation involving the Gulf Stream. The way to cool the surface layer at lower latitudes is to pump up cooler water from the underlying layer, creating a feedback, a kind of convection process, with surface water cooling and sinking while deeper water warms and rises. The upwelling water would be rich in nutrients and encourage plankton to bloom, drawing carbon dioxide out of the air and reducing the greenhouse effect. This ocean fertilization process already occurs naturally when the sea is stirred up by hurricanes.

Lovelock has discussed the idea with an entrepreneur who has a background in physics. They reckon that the process could be made to work using tubes of polypropylene about 10 metres in diameter and 200 metres long, weighted at one end so that they float vertically with their tops just below the surface of the sea. A little pumping would be required to start the process working, but even that could be provided free by using wave action to operate flaps at the ends of the tubes. The cost would be small, and the whole system could even be profitable. Lovelock suggests that the Gulf of Mexico would be the ideal place for such a system, because it would cool the surface layer and thereby reduce the strength of hurricanes, it would improve the fisheries by fertilizing the sea, and there are plenty of old oil rigs there to use as platforms for the pipes to be hung from.

He is just as enthusiastic about the Cool Earth project, which has been set up to use donations from individuals and organizations to buy rainforest and leave it undisturbed. 'This,' he says, 'is Gaian thinking – systems thinking, not thinking like an individual human being. It makes much more sense than misguided efforts at planting new forests. The ones we've got have evolved to do the job better than anything we might manage.' His optimism shines through also at a more personal level. A man holding the views expressed in *Revenge of Gaia* and with nine grandchildren ranging in age from sub-teens to thirties might be expected to be a little gloomy about their prospects, but 'I see them as among the survivors; the young see disasters as potential adventures.'

So how does Lovelock see the future for people at large in his own ninetieth year of life? He reminds us that the ultimate cause of the problem is that there are too many people – six or seven times too many people – on Earth today, and that this will have to come down whatever the success of any of the ideas outlined here. He decries the amount of money and effort that goes in to trying to prolong the human lifetime, which he sees as having 'a natural biological limit of about a hundred years'. He is much more approving of the hospice movement, providing people in affluent countries with the opportunity for a timely and dignified death. 'The past twenty years have been the happiest of my life, and I'm ready to accept the end when it comes. I stand by what I said at the end of *Homage*.' Life was present on

Earth about four billion years ago, and by Lovelock's reckoning the inexorable warming of the Sun will make life on Earth impossible in about a billion years time. At the end of *Homage*, written when he was nearly eighty, Lovelock pointed out that both he and Gaia were about four fifths of the way through their likely lifespan. 'It is comforting,' he concluded, 'to know that my destiny is to merge with the chemistry of our living planet.'

But he isn't quite ready to do this yet. Richard Branson, the founder of the Virgin business empire, has offered a prize of $25 million for anyone who comes up with a workable technological fix for global warming. Lovelock, naturally, has been asked to be one of the judges of the entries for the competition. When he met Branson to discuss the arrangements, he also mentioned his fascination with space, and how much he envied the people who will soon be able to fly above the Earth's atmosphere in the Virgin Galactic spaceplane. Branson promptly offered him a flight in 2009, when Lovelock will be 90. 'Of course, I accepted,' he told us in the summer of 2007. 'It's my boyhood fantasy come true. And if anything goes wrong, what a way to go!'

CODA
Making an Invention

Like many people, we were curious about how a creative person like Jim Lovelock comes up with his ideas – what does it take to be an inventor? Here is the answer, in his own words.

There are many different ways of inventing. For me there is usually, but not always, the requirement of a need.

For example, in 1946 my boss Robert Bourdillon told me that a sensitive anemometer was needed (see Chapter 4). He explained that the feeling of comfort in a room depends among other things on the rate of air movement, and this could be important at velocities as low as 0.5 cm/second.

He went on to say that the available hot-wire anemometer was too insensitive for air movement as slow as this and consequently physiologists used a strange device called a katathermometer. It was like any ordinary thermometer except that the bulb was a glass vessel of about 100 ml capacity filled with a weak solution of dye in alcohol. The average air movement of a room was measured by warming the katathermometer to 36°C and observing how long it took to fall five degrees in temperature. Each measurement took a long time. He needed something faster than that, with a response in seconds or less.

My first thought was, what physical processes have an electrical signal directly transduceable from air movement? Two came to mind. First, the speed of sound waves in air depends on the velocity of the air as well as on temperature. The second idea was that the drift of ions in air is slow (1 cm/sec per volt/cm), so monitoring the movement of ions would give a good guide to the speed of air currents. After that, to make prototypes of either kind of working anemometer was just a matter of putting the hardware together. For the ionization anemometer it was a bit hairy as it meant scraping the radioactive paint from the dials of old aircraft, to get the radioactive isotopes needed for the ionization process; but apart from that it was very straightforward. The sonic anemometer was safer, but more difficult because no sensitive microphones for frequencies above 20 kHz, in the range I needed, existed in those days, so I had to use a 200 watt source of sound at 21 kHz. Sensitivity was proportional to frequency.

Both anemometers worked at the desired sensitivity, but the ionization device was much the more practical then. The results were described in a paper with my first graduate student in *Review of Scientific Instruments*, in 1948. Now, of course, the sonic anemometer is widely used and employs small efficient ceramic transducers. The ion anemometer fell out of favour when the fear of radiation became ubiquitous.

On other occasions, inventions flash upon that inward eye and cause the loss of early morning sleep. Daisyworld was one of these. Such inventions are clearly a product of the unconscious mind at work, and as such are largely inexplicable.

James Lovelock
Coombe Mill
March 2008

Acknowledgements

We are grateful to many scientists who gave up their time to discuss the Gaia concept, global warming and the work of Jim Lovelock with us, either face to face or electronically. These include: Richard Dawkins, James Fleming, James Hansen, Lynn Margulis, Ken Muldrew, Fred Pearce, Chris Rapley, Sherwood Rowland, Stephen Schneider, Richard Somerville, Andreas Sputtek, Sir Crispin Tickell, Scott Turner, Andrew Watson, Spencer Weart and Karl Woodcock.

Several libraries and research institutions also provided invaluable help and access to their collections: the Bodleian Library; Cambridge University Library; the Royal Astronomical Society; the Royal Geographical Society; the Royal Institution; the Royal Society; Trinity College, Dublin. As ever, the University of Sussex provided us with a base from which to work, and the Alfred C. Munger Foundation provided a contribution to our travel and other expenses.

Parts of Chapter 6 are adapted from *The Hole in the Sky*, by John Gribbin (Corgi, 1988).

Sources and Further Reading

(The editions referred to below are the ones we have had access to; in some cases these are reprints, or translations, of the first editions.)

Louis Agassiz, *Études sur les glaciers*, published privately by the author, Neuchâtel, 1840.

Svante Arrhenius, *Worlds in the Making*, Harper, New York, 1908.

Robert Berner, *The Phanerozoic Carbon Cycle*, Oxford UP, 2004.

Rachel Carson, *Silent Spring*, Houghton Mifflin, New York, 1962 (still in print as a Penguin Modern Classic, Penguin, London, 2000).

John Cox, *Climate Crash: Abrupt Climate Change and What it Means for Our Future*, Joseph Henry Press, Washington, 2005.

James Croll, *Climate and Time in their Geological Relations*, Daldy, Isbister & Co., London, 1875.

James Croll, *Discussion on Climate and Cosmology*, Appleton, New York, 1886.

Charles Darwin, *The Formation of Vegetable Mould, Through the Action of Worms with Observations on Their Habits*, John Murray, London, 1881.

Richard Dawkins, *The Extended Phenotype*, Oxford UP, revised edition, 1989 (first published 1982).

Lydia Dotto and Harold Schiff, *The Ozone War*, Doubleday, New York, 1978.

Arthur Eve and Clarence Cressey, *The Life and Work of John Tyndall*, Macmillan, London, 1945.

James Fleming, *The Callendar Effect*, American Meteorological Society, Boston, 2006.

Joseph Fourier, *Oeuvres de Fourier*, Volume Two (edited by Gaston Darboux), Gauthier-Villars, Paris, 1890.

John Gribbin, *Hothouse Earth*, Bantam, London, 1990.

John Gribbin, *Science: A History*, Allen Lane, London, 2002 (US title: *The Scientists*).

John Gribbin, *Deep Simplicity*, Allen Lane, London, 2004.

J. B. S. Haldane, *My Friend Mr Leakey*, Puffin, London, 1944.

Alexander von Humboldt, *Personal Narrative*, trans. Jason Wilson, Penguin, London, 1995 (originally published in three volumes in French between 1814 and 1825).

Alexander von Humboldt, *Cosmos*, trans. E. C. Otté, Johns Hopkins UP, 1997 (originally published in German in five volumes between 1845 and 1861).

John Imbrie and Katherine Palmer Imbrie, *Ice Ages: Solving the Mystery*, revised edition, Harvard UP, 1986.

Robin Jones and Tom Wigley (eds.), *Ozone Depletion: Health and Environmental Consequences*, Wiley, Chichester, 1989.

Gerard Kuiper (ed.), *The Earth as a Planet*, University of Chicago Press, Chicago, 1954.

Lee Kump, James Kasting and Robert Crane, *The Earth System*, Prentice Hall, New Jersey, 1999.

Emmanuel Le Roy Ladurie, *Histoire du climat depuis l'an mil*, Flammarion, Paris, 1967.

Hubert Lamb, *Climate, History and the Modern World*, Routledge, London, 1955.

Richard Leakey and Roger Lewin, *The Sixth Extinction*, Anchor, New York, 1996.

Alfred Lotka, *Elements of Mathematical Biology*, Dover, New York, 1956 (reprint of a book originally published in 1925 as *Elements of Physical Biology*).

James Lovelock, *Gaia: A New Look at Life on Earth*, revised edition, Oxford UP, 2000 (first published 1979).

James Lovelock, *The Ages of Gaia*, revised edition, Oxford UP, 2000 (first published 1988).

James Lovelock, *Gaia: The Practical Science of Planetary Medicine*, Gaia Books, London, 1991.

James Lovelock, *Homage to Gaia*, Oxford UP, 2000.

James Lovelock, *The Revenge of Gaia*, Allen Lane, London, 2006.

Mark Lynas, *Six Degrees*, Fourth Estate, London, 2007.

Lynn Margulis, *The Symbiotic Planet*, Weidenfeld & Nicolson, London, 1998.

Lynn Margulis and Lorraine Olendzenski (eds.), *Environmental Evolution*, MIT Press, 1992.

Lynn Margulis and Dorion Sagan, *Microcosmos*, Summit, New York, 1986.

Lynn Margulis and Dorion Sagan, *What is Life?*, Weidenfeld & Nicolson, London, 1995.

James Martin, *The Meaning of the 21st Century*, Eden Project Books, London, 2006.

Mary Midgley, *Gaia: The Next Big Idea*, Demos, London, 2001.

Milutin Milankovitch, *Théorie mathématique des phénomènes thermiques produits par la radiation solaire*, Gauthier-Villars, Paris, 1920.

Milutin Milankovitch, *Kanon der Erdbestrahlung und seine Anwendung auf das Eiszeitenproblem*, Royal Serbian Academy, Belgrade, 1941.

Fred Pearce, *The Last Generation*, Eden Project Books, London, 2006.

Sharon Roan, *Ozone Crisis*, Wiley, New York, 1989.

Aaron Sachs, *The Humboldt Current*, Oxford UP, 2007.

Eric Schneider and Dorion Sagan, *Into the Cool: Energy Flow, Thermodynamics and Life*, University of Chicago Press, 2005.

Stephen Schneider, *Global Warming: Are We Entering the Greenhouse Century?*, Sierra Club Books, San Francisco, 1989.

Stephen Schneider, *Laboratory Earth*, Basic Books, New York, 1996.

Stephen Schneider and Penelope Boston (eds.), *Scientists on Gaia*, MIT Press, 1991.

Stephen Schneider and Randi Londer, *The Coevolution of Climate & Life*, Sierra Club, San Francisco, 1984.

Stephen Schneider and Lynne Mesirow, *The Genesis Strategy: Climate and Global Survival*, Plenum Press, New York, 1976.

Stephen Schneider, James Miller, Eileen Crist and Penelope Boston (eds.), *Scientists Debate Gaia*, MIT Press, 2004.

Erwin Schrödinger, *What is Life?*, Cambridge UP, 1967 (originally published in 1944; this edition combined in one volume with the same author's *Mind and Matter*, from 1958).

Eduard Suess, *Die Entstehung der Alpen*, Braunmüller, Vienna, 1875.

E. T. Sundquist and W. S. Broecker (eds.), *The Carbon Cycle and Atmospheric* CO_2, American Geophysical Union, Washington, 1985.

Lewis Thomas, *The Lives of a Cell*, Viking, New York, 1974.

Scott Turner, *The Tinkerer's Accomplice*, Harvard UP, 2007.

John Tyndall, *Contributions to Molecular Physics in the Domain of Radiant Heat*, Appleton, New York, 1873.

Vladimir Vernadskii, *The Biosphere*, annotated edition, Copernicus, New York, 1997 (translation of *Biosfera*, Nauka, Leningrad, 1926). See also the French edition, *La Biosphère*, Felix Alcan, Paris, 1929.

Tyler Volk, *Gaia's Body*, Copernicus, New York, 1998.

Spencer Weart, *The Discovery of Global Warming*, Harvard UP, 2003 (the online hypertext linked with this book, at *http://www.aip.org/history/climate*, includes an extensive list of references to scientific publications relevant to the story of global warming).

Robert Weinberg, *One Renegade Cell*, Basic Books, New York, 1988.

Jonathan Weiner, *The Next One Hundred Years: Shaping the Fate of Our Living Earth*, Bantam, New York, 1990.

Edward O. Wilson, *The Diversity of Life*, Harvard UP, 1992.

Edward O. Wilson, *Consilience*, Little, Brown, London, 1998.

Edward O. Wilson, *The Future of Life*, Knopf, New York, 2002.

J. Z. Young, *An Introduction to the Study of Man*, Clarendon Press, Oxford, 1971.

Index